Sikorsky VS-44
Flying Boat

Harry E. Pember

Table of Contents

Dedication:

To the men and women who designed, built, flew, and restored the last of the Sikorsky flying boats. Also, to my wife Bunny and to my three sons, D.J., Jon, and Chris, who spent many Saturday mornings "flying" *Excambian* while Dad interviewed project volunteers. D.J. probably has as many "Pilot-in-Command" hours on the VS-44A as Dick Probert.

Cover: VS-44A *Excambian* in Navy camouflage with American Export markings.

© 1998 by Harry E. Pember and Flying Machines Press

Published in the United States by Flying Machines Press, 35 Chelsea Street, Stratford, Connecticut, 06497

Book and cover design, layout, and typesetting by John W. Herris.

Cover painting by Joseph Keogan, courtesy of the Igor I. Sikorsky Historical Archives.

Color aircraft illustrations and scale drawings by Juanita Franzi.

Digital scanning and image editing by Gretchen H. Herris and John W. Herris.

Text edited by John W. Herris.

Printed and Bound in Korea.

Publisher's Cataloging-in-Publication Data

Pember, Harry E., 1949–
Sikorsky VS-44 Flying Boat / Harry Pember
     p.   cm.
Includes bibliographical references.
ISBN 1-891268-02-3 (alk. paper)
1. Airplanes, Military—American—History.
2. World War, 1939–1945—Aerial operations, American.
3. World War, 1939–1945—Equipment.
UG1245.F8D38   1998
623.7'461'094409041—dc21   98-xxxxx
     CIP

Flying Machines Press

For the latest information on our books and posters, visit our web page at:
www.flying-machines.com

Introduction

In 1918, Igor Ivanovich Sikorsky was at the height of a very successful career designing and building large bombers and scout aircraft for the Imperial Russian Air Service. From the very earliest days of World War I, Sikorsky's giant four-engined bombers, the Il'ya Muromets series,[1] had been quite effective in creating chaos for their German adversaries. His little S-16 scouts were a match for any enemy aircraft that they encountered over the Eastern Front. Arguably, Sikorsky had become Russia's most prominent aircraft designer. However, his work was interrupted by the Russian Revolution. As a supporter of the Czar and an obvious threat to the Bolsheviks, he was forced to flee his native land. He traveled first to London and then on to France, where he was commissioned to design a heavy bomber for the French government. After the signing of the armistice ended WWI, the French canceled his contract. Sikorsky then chose to immigrate to the United States, where he felt that he had the best opportunity to pursue his career in aviation. On his passport, Sikorsky listed his intended occupation as "Engaging in the business of designing and constructing large aircraft." He privately hoped to build the world's largest airplane.

It took Sikorsky nearly three years to assemble the financial backing to embark on a new aircraft project. Finally, in 1923 Sikorsky and his team built one of the world's first all-metal transport planes, the S-29A. In the years that followed, several other landplane and amphibion[2] designs were successfully built, but unfortunately, in the mid 1920s the market for new airplanes was

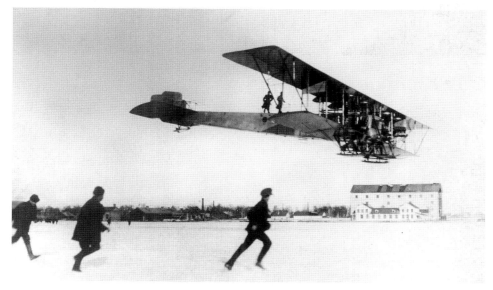

Top: The S-16, one of Igor Sikorsky's favorite designs. This particular aircraft, serial number 155, was the second S-16 manufactured by the Russo-Baltic Railroad Car Company (R-BVZ). It was powered by an 80hp Gnome engine. (IISHA)

Above: Sikorsky's most famous Russian design, the Il'ya Muromets four-engine bomber and long-range reconnaissance plane. (IISHA)

Left: The success of the S-38 Amphibion provided notoriety and financial security for the Sikorsky Manufacturing Company. This early model is seen flying over the Sikorsky Plant at College Point, Long Island, New York. (IISHA)

The S-40 *American Clipper*. A total of three were built for Pan American Airways. Introduced in 1931, they were used to establish passenger travel to South America. During World War Two , they were utilized by the U.S. Army as flight trainers and, after long, incident-free operational lives were ultimately scrapped after the war. (IISHA)

The S-42 *Pan American Clipper* over the yet to be completed Golden Gate Bridge circa 1935. (IISHA)

Igor Sikorsky, Charles Lindbergh, PAA chief engineer Andre Priester, Sikorsky Chief Pilot Boris Sergievsky, and PAA chief pilot Ed Musick at the seaplane ramp at Stratford, Conn., after the S-42 successful world record flight. (IISHA)

very limited. World War I aircraft such as the JN-4 "Jenny" could still be purchased for a few hundred dollars. Air passenger service had barely emerged as a viable mode of travel. Public skepticism of air travel was still at a high level. The business of aviation was, for the most part, at a standstill. The Sikorsky Manufacturing Company was barely keeping its doors open.

However, in the spring of 1927, the flight of the "Lone Eagle," Charles Lindbergh, from Roosevelt Field on Long Island non-stop to Paris, France, captured the hearts and imagination of Americans everywhere. Interest in all-things aviation blossomed overnight. Americans suddenly became comfortable with the idea of air travel and airports began to spring up in nearly every major city. Igor Sikorsky sensed this enthusiasm and asked his associates at his plant in College Point, Long Island, to accelerate the development of their latest design, the twin-engine passenger carrying amphibian, designated the S-38.

The S-38 first flew in late May, 1928. Its performance proved to be superior to any amphibian then flying anywhere in the world. Powered by two P&W (Pratt & Whitney) 420hp "Wasps," it was capable of carrying ten people at top speeds of

130 mph, cruising at 100 mph, climbing with full load at 1,000 fpm (feet per minute), and most importantly it had the ability to continue in level flight at full load on one engine. The first series was sold out quickly, most going to the U.S. Navy or Pan American Airways. A second series of ten were started and sold from the production line. The Sikorsky Company found itself with more orders than they could produce in their Long Island facilities.

In 1929 an S-38 was used by Colonel Lindbergh to inaugurate airmail service between the United States and the Panama Canal Zone. The success of the S-38 attracted new investors which enabled the company to undergo a $5,000,000 re-capitalization as the Sikorsky Aviation Company. As many of these investors were from New England, Igor Sikorsky was urged to build his plant in Connecticut. He found an ideal location, on the Housatonic River in Stratford, where he constructed the new, larger Sikorsky factory. Later that same year, the Sikorsky Aviation Company became a subsidiary of The United Aircraft Corporation.

The S-38 was used by literally dozens of airlines, the military services, and many private individuals as well.

Prominent individuals flying the S-38 included Charles Lindbergh, explorers Martin and Osa Johnson, and the Prince of Wales. Between 1928 and the early 1930s, some 110 were built and sold.

In late 1929, Pan American Airways approached Sikorsky with a proposal to built a much larger flying boat. Collaborating with Pan American's chief engineer, Andre Priester, and Pan American consultant Lindbergh, Sikorsky concluded that the best approach would be to construct a new airplane based in large part on the design of the venerable and proven S-38. Newer technologies such as the fully canti-levered wing were then available, but Sikorsky convinced his associates that his approach was the safest and would guarantee a rapid and successful expansion to the Pan American fleet. In 1930, Pan American issued Sikorsky a contract to build what was to be the largest airplane yet built in the United States. The S-40, when completed in 1931, carried 40 passengers in great comfort over distances of 500 miles at cruise speeds of 115 mph. In October 1931, the first of these airliners flew to Washington, D.C., where it was christened *American Clipper* by the wife of President Hoover. It was to become the

The S-43 on step during acceptance testing on the Housatonic River, Stratford, Conn. (IISHA)

first of a long line of Pan American Clippers. The *American Clipper* departed from Miami on its maiden voyage on November 19, 1931, piloted by Pan American's chief pilot Basil Rowe, with Charles Lindbergh and Igor Sikorsky as co-pilots. A total of three S-40s were built, all for Pan Am. They flew, virtually without incident, for millions of miles during the 1930s exploring new passenger routes in the Caribbean and South America, were used as navigation trainers during World War II, and eventually were retired and scrapped between 1944 and 1946.

The next Sikorsky airplane built was the S-41, which was essentially a scaled-up S-38 that held 14 passengers. Other than size, the most significant difference between the two aircraft was that the S-38 was a sesquiplane (a biplane with the lower wing having a much smaller area than the upper wing) and the S-41 was a monoplane. Only seven S-41s were built.

Sikorsky's next aircraft, the S-42, was conceived in discussions which took place between Igor Sikorsky, Charles Lindbergh, and Basil Rowe during the maiden voyage of the S-40. Basically, they conceived the challenge to build an aircraft using the latest technology, which would be a true trans-oceanic airliner, capable of carrying enough fuel

to fly 2,500 miles non-stop against a 30 mph headwind at a cruising speed of 150 mph.

The S-42, which first flew in 1934, broke with tradition by doubling the existing wing loading from 15 to 30 pounds per square foot, twice what was then considered the maximum allowable for commercial, passenger-carrying aircraft. The S-42 was powered by four P&W 750hp engines and had the first commercial installation of the new variable-pitch propellers designed by Hamilton-Standard. The S-42 also featured a high aspect-ratio wing with an advanced three-quarter-span flap which aided take-off and provided a conservative 65 mph landing speed. This new technology produced a top speed of 188 mph, cruise speed of 160 mph, a range of 1,200 miles with a payload of 7,000 pounds, and a range of 3,000 miles if the payload was reduced to 1,500 pounds. Shortly after its first test flights were completed, it was used to establish a series of new world records, which put the United States into the lead as far as total records were concerned; holding 17 to France's 16 in the summer of 1934. Ten of the clippers were delivered to Pan Am and were utilized to pioneer scheduled air service across both the Atlantic and Pacific Oceans.

The End Of An Alliance

Clearly, by the mid-1930s the name Sikorsky had become synonymous with the term "flying boat" and it seemed that the fortunes of Sikorsky and Pan American would remain tightly intertwined. Surprisingly, when Pan American established a requirement for a "giant" flying boat that could span both oceans while carrying 50 or more passengers, the Sikorsky design lost out to the Martin M-130. The long relationship between the two companies was effectively and immediately dissolved. However, Igor Sikorsky was still convinced of the practicality of the flying boat and was destined to design and produce one more history-making "boat," the S-44, which was originally designed to fill a need of the U.S. Navy and went on to set several records as a commercial transport—ironically enough, for Pan American Airway's number one competitor.

XPBS-1 "Flying Dreadnought" (Sikorsky S-44)

In the early months of 1935 the U.S. Navy Bureau of Aeronautics (BuAer) undertook an investigation to determine the feasibility of designing a seaplane or flying boat which would have much greater performance and range than the currently operational PBY Catalina. The vast distances between island bases in the Pacific and the area requiring air patrol dictated the need for a new, more capable aircraft. Addressing this issue, BuAer representative Rear Admiral Joseph Reeves, USN, discussed the Navy's needs with Eugene E. Wilson, former naval aviator and senior vice-president of United Aircraft Corporation. Wilson informed Reeves that United Aircraft had already conducted an independent study which indicated that a new breed of larger flying boats could best solve the Navy's problem. Reeves was convinced, and consequently BuAer released Design Proposal #137 which formally requested bids on a new flying boat for the U.S. Navy. In short, the aircraft was to function as a patrol bomber having a range of 3,450 miles.[3] Additional requirements included a maximum speed of 200 miles per hour, four machine-gun turrets, and accommodations for a crew of six.

Although several aviation companies responded to BuAer's request for designs, the proposal submitted by Sikorsky Aircraft was the only one immediately acceptable to the Navy Department. Consequently, on June 25, 1936, Sikorsky Aircraft was issued a contract to build a single prototype patrol bomber. The Navy officially designated this aircraft the XPBS-1; however, the Sikorsky Aircraft Company's internal designation for this design was S-44 (the 44th aircraft model to be designed and built by Sikorsky).[4] In detail, the Sikorsky design specified an all-metal construction, single high-mounted, full canti-lever wing, powered by four Pratt & Whitney Twin Wasp R-1535, 700hp radial engines and Hamilton Standard constant-speed, $10\frac{1}{2}$ foot diameter three-bladed propellers. The aircraft's armament would include one 0.50-cal. machine gun in both the bow and tail turrets and one 0.30-cal. machine gun in each of the two center gun turrets. Positioning a gun turret at the tail section of the aircraft made the XPBS-1 the first American airplane to utilize this means of defense. However, considering Igor Sikorsky had designed the world's first tail gun emplacement for use on an aircraft, this was a natural development for the XPBS-1.[5] In addition to defensive weapons, the aircraft was also designed to carry a bomb load of 4,000 lbs. for offensive purposes.

In a sense, the XPBS-1 was a true flying boat with provisions for carrying beaching gear to facilitate maintenance. Other innovations in this aircraft were the installation of a 110-volt electrical system, a gasoline generator which supplied power to the flaps, anchor winch, radio, lights, bomb controls, and galley. In addition, the XPBS-1 also had a

Mr. Sikorsky in front of a painting of the XPBS-1 "Flying Dreadnought" flying over the aircraft carrier *Saratoga*. (IISHA)

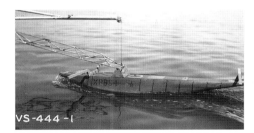

One of several water test models of the XPBS-1 used to predict performance of the hull during take-off and landing. (IISHA)

The "Flying Dreadnought" resting on her beaching gear at the Sikorsky ramp. (IISHA)

VS-438 XPBS-1 WRINKLES IN RIGHT HAND L.E. & NACELLE COWLINGS 8-11-39

A close-up of the two starboard XPBS-1 engines. (IISHA)

The structure and framing of the XPBS-1 pilot's cabin.

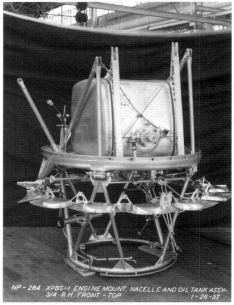

NP-284 XPBS-1 ENGINE MOUNT, NACELLE AND OIL TANK ASSY- 3/4 R.H. FRONT - TOP   1-26-37

The mount, nacelle, and oil tank of a XPBS-1 engine.

complete telephone system which made communication possible throughout the ship. The ailerons and flaps were of the same technology utilized on both the F4U-1 Corsair and OS2U-1 Kingfisher. The hull design of the XPBS-1 was based in large part on Sikorsky's experience with the S-42. The hydrodynamic test hull was tested in the Washington Navy Yard Model Basin and the wind tunnel model was tested at NACA and MIT.

The gross weight of the XPBS-1 was 47,142 pounds in the bomber configuration and 49,059 pounds in the patrol configuration.

Supporters of the XPBS-1 envisioned the aircraft as an airborne version of the waterborne "Dreadnought" battleship. Therefore, the XPBS-1 became unofficially, but affectionately, known throughout the U.S. Navy as the "Flying Dreadnought."

The XPBS-1 during flight tests in 1937 over southern Connecticut. (IISHA)

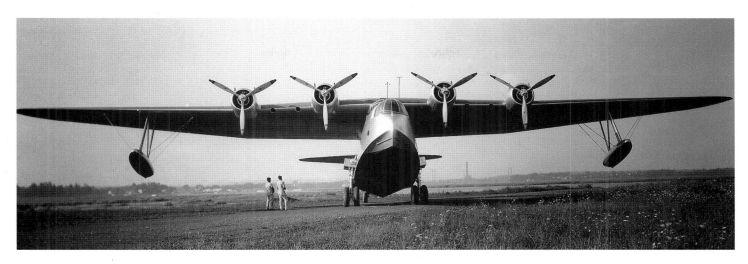

The XPBS-1 on the Sikorsky ramp. The alignment of the propellers is indicative of the "nothing less than perfect" attitude of the Sikorsky crew. (IISHA)

An aerial photo of the XPBS-1 taxiing up to the Sikorsky seaplane ramp during initial flight tests in 1937. The S-44 hangar is upper center in the photo and the cylindrical structure behind it is the vertical wind tunnel. The ramp and all of the buildings in this photograph exist today. (IISHA)

One month after the Sikorsky Aircraft Co. received its contract to build the XPBS-1 prototype, BuAer awarded Consolidated-Vultee Aircraft Co. a contract to develop a second flying boat design intended primarily to compete with the XPBS-1. The Consolidated design was designated the XPB2Y-1 Coronado.

The U.S. Navy decreed that both the Coronado and Dreadnought were to be powered by the newly developed Pratt & Whitney R-1830-68 engines rated at 1,050hp. Twelve-foot diameter constant-speed, hydro-controllable three-bladed

Above: The "Flying Dreadnought" over Lordship, Conn. The Remington Gun Club is under the starboard wing of the aircraft. (IISHA)

Right: The XPBS-1 water taxiing at the mouth of the Housatonic River. (IISHA)

propellers were required as well for both aircraft. These upgrades insured better performance for the XPBS-1, including increasing the bomb load by an additional 8,000 lbs!

In a 1937 news release regarding the VS-44, the U.S. Navy stated that "The construction of this plane was undertaken by the Navy Department in its efforts to explore the value of large flying boats in national defense, as for years it (the U.S. Navy) has sponsored their development by well regulated experiments. This flying boat represents one of the most powerful bombing planes in the U.S., having a military load carrying capacity comparable to that of any known existing airplane. It will also have the usual long range demanded of Navy patrol bombers, as demonstrated by the Twin Wasp-powered PBY-1 patrol bombers in the recent non-stop flights from the Naval Air Station, San Diego, to the Fleet Air Bases at Pearl Harbor, T.H., and Coco Bolo, Canal Zone. It will afford the Navy Department an opportunity to compare the relative value as a national defense weapon, both from a tactical and an engineering standpoint, the large four-engine flying boats versus the smaller two-engine type flying boats."

The Navy further described the aircraft as having "every known approved device for safety and ease of

Above: A view of the XPBS-1 at anchor in the Housatonic River. Note the tail gun emplacement and large cargo doors. (IISHA)

Right: This photo, documenting repairs to the hull which was damaged by hitting floating debris during take-off, provides an excellent illustration of the construction of the XPBS-1 hull, including bulkheads and longitudinal stringers. (IISHA)

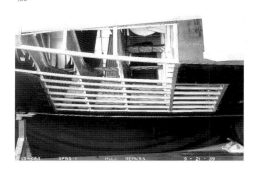

Another view of the XPBS-1 on step. (IISHA)

operation incorporated in this airplane. Ailerons and flaps are of all metal construction, fabric covered, with the full trailing edge flap permitting rapid take-off and slow landing speeds. A complete radio compartment is installed with radio equipment comparable to that on a modern destroyer. Soundproof throughout and equipped with commodious living accommodations for the crew with a mechanics workshop, galley with electric stove, water distiller, and dry ice refrigerator, sustained operation of the XPBS-1 is possible and the physical endurance of the personnel increased. Civil aviation will benefit greatly with the development of this new airplane, when it is realized that by adapting its large weight carrying capacity to the carrying of passengers, our commercial airlines will have the most modern flying boat equipment in existence."[6]

The first flight of the XPBS-1 took place on August 13, 1937, in the Housatonic River near the Sikorsky plant in Stratford, Conn. On board were Edmund T. Allen, pilot, and Clifford Swartz, co-pilot. In the cabin rode a select crew of Navy officials and Sikorsky engineers. The Navy was represented by Lt.Cdr. Z.C. Sanborne and Lt.Cdr. J.C. Duerfeldt, while Michael Gluhareff, Knute Henrichsen, A. Zuiger, and Anthony Fokker, Jr., made up the Sikorsky Engineering contingent. Dean Taylor and Eddie Lund served as mechanics. Bill Heiden, of Sikorsky, was the crew chief. On shore by the Stratford Lighthouse, watching closely the performance of the big boat, was Lt. Rhea S. Taylor, USN, who as Inspector of Naval Aircraft, had keenly observed the development and construction of the XPBS-1 during the previous two-year period.

The XPBS-1 in the massive Sikorsky Hangar next to the world's first practical helicopter, the VS-300. The helicopter would make its first flight one week later. (IISHA)

An excellent view of the tail section of the XPBS-1 including the rear gun emplacement.

The XPBS-1 on step—exact location uncertain—perhaps San Francisco or Alameda. (IISHA)

Taylor was quickly relieved; the XPBS-1 took off smoothly and for the first hour raced down the Connecticut shoreline to Norwalk Harbor, up the shoreline to Old Saybrook, and back down to Stratford again with speeds nearing 200 mph. Then Allen throttled back the roaring Twin Wasps and the Navy's newest aerial weapon settled easily to a perfect landing, skimming along the surface of the bay with a plume of white water boiling in its wake.

Between August 13 and October 7, 1937, Sikorsky aircrews flew 26.9 hours of evaluation flight tests and established the aircraft's top speed of 225 mph at 10,000ft., a stall speed of 62 mph, and a take-off time of 30 seconds. Initial climb rate was determined to be 640 feet per minute. Service ceiling was 23,100 feet and normal and maximum ranges were 3,170 miles and 4,545 miles, respectively. During testing it was noted that the XPBS-1 exhibited some undesirable stall characteristics during landing. Many of these characteristics were later discovered to be associated with the downwash of the wing on the horizontal tail. By the conclusion of formal testing, virtually all critical performance parameters exceeded contract specifications.

As an example of how this huge aircraft captured the imagination of the American public, it is interesting to note that shortly after the XPBS-1 began flight testing, the famous cartoonist Zack Mosely made it the airplane of choice for his cartoon hero "Smilin' Jack."

Sikorsky delivered the XPBS-1 proto-type to the Norfolk Naval Air Station on October 12, 1937. The aircraft was in turn assigned to Navy PatWing 5 which operated out of NAS Norfolk, Virginia. On October 28, formal Navy acceptance testing began. Engine problems caused significant delays in the testing, but the tests were completed on July 5 with the Navy logging a total of 37 flights and 53.5 flight hours. Overall, Navy pilots were pleased with the performance of the XPBS-1. In fact, only one minor complaint was noted. Control forces, especially relative to the large rudder, were unacceptably high at speeds reaching cruise and above. Once testing was concluded, the aircraft was returned to the Sikorsky plant where appropriate modifications were performed. While undergoing these modifications, the XPBS-1 lost the production contract to the Coronado. This was ironic, considering, that the Navy evaluation crews concluded in their final report that "the XPBS-1 had bombing characteristics superior to any other airplane then in the Navy inventory and is recommended as a suitable type for service use as a patrol-bomber flying boat."

Reportedly, the final production decision was based primarily on price. That premise is difficult to prove because very little pricing information is available for either model today. It was also known that the United Aircraft Corporation was most concerned with supplying the Hamilton Standard propellers and Pratt & Whitney engines at a designated price, regardless of the winner of the airframe competition. The corporation was, evidently, unwilling to reduce the airframe price. Beyond the prototype, no additional XPBS-1 models were produced and, sadly, the prototype spent the balance of her service life performing mostly routine duties as a passenger/cargo aircraft. In this role, one of the aircraft's earliest assignments was serving the newly established U.S. Neutrality Patrol, which operated out of bases along the Atlantic coast such as Norfolk, Newport, Charleston, Key West, San Juan, Trinidad, Bermuda, and Argentina. Later, during October 1941, the aircraft carried numerous congressmen to and from Bermuda to participate in discussions of wartime strategy with allied leaders. In 1942 the aircraft was placed under the command of Lt. Norman Miller, USN, who also commanded the aircraft on a very strenuous trip to deliver spare parts for U.S. Navy PBYs operating in the war-torn South Pacific. Mission completed, Lt. Miller returned the aircraft to San Diego, California, where it underwent scheduled overhaul and repair.[7]

The XPBS-1 was subsequently assigned to Air Transport Squadron Two

Above: The XPBS-1 at Norfolk NAS during acceptance testing. (IISHA)

Right: The XPBS-1 at Norfolk NAS during acceptance testing. (IISHA)

(UR-2) located at NAS Alameda, Cal. On June 30, 1942, while returning from Pearl Harbor, the aircraft sustained serious damage when it struck a large floating log on landing in San Francisco Bay. The aircraft was totally destroyed and was subsequently stricken from the inventory. Fortunately, there were no fatalities or injuries reported in the incident, an important fact, considering Admiral Chester Nimitz was reportedly on board at the time of the crash. The total flight time on the aircraft when it was stricken from the Navy inventory was 1,365.7 hours.

VS-44A Design, Construction, and Operations

American Export Airlines

In 1937, when the Sikorsky Aircraft Company was busy building the XPBS-1 prototype, it was becoming obvious to Juan Trippe that his company, Pan American Airlines, was beginning to lose its monopoly as the unofficial United States flag carrier. His concerns were all too real! That same year, the American Export Shipping Line established a subsidiary, American Export Airlines (AEA), and aggressively petitioned Congress for permission to explore air routes across the Atlantic and throughout the Caribbean. To make matters worse for Trippe, AEA had been purchased by a large syndicate which included, among others, Robert Lehman of the renowned banking firm of Lehman Brothers. Unfortunately for Trippe, Lehman had been a trusted confidant while serving on the Board of Directors of Pan American Airlines. As other key Pan American employees joined AEA, it became evident that PAA was now in a very competitive market, especially relative to the most valuable Trans-Atlantic passenger routes.

Vought-Sikorsky Aircraft

While PAA and AEA competed for market share, in 1937 the United Aircraft Corporation combined operations of the Chance Vought Division with those of Sikorsky Aircraft, forming the Vought-Sikorsky Division. The new entity was consolidated at the Sikorsky Aircraft plant in Stratford, Conn. This was done primarily to allow the bustling Pratt & Whitney Division room to expand into the Vought factory located in Hartford, Conn. UAC management felt this merger would increase the overall operating efficiency of the new entity, and to ensure successful accomplishment of this goal, UAC appointed a management team which included several very notable individuals. Mr. C.J. McCarthy was named General Manager, Igor I. Sikorsky, Engineering Manager, Rex Beisel, Chief Engineer, and Michael E. Gluhareff, Chief of Design. Although both "divisions" continued to concentrate on their particular product lines, there was a significant amount of "cross pollination," especially in the engineering department. At that time, the internal company designation for the XPBS-1 was changed from the S-44 to the VS-44 (denoting a Vought-Sikorsky design).

In addition to the VS-44A flying boat, the Vought-Sikorsky Company was busy with several military programs; the

VS-44A #41880 before joining of center section (September 1941). Note less complete pilot's cabin of #41881 at bottom of photo. (IISHA)

THE VS-44A ROLLS OUT ~~ON SCHEDULE~~ ~~AHEAD OF SCHEDULE~~

A 1941 cartoon illustrated the "hoopla" surrounding the roll-out of the first VS-44A. (IISHA)

OS2U Kingfisher, The SB2U Vindicator dive bomber, the S-43 twin-engine amphibian, the VS-173 "flying pancake," and of course the F4U-1 Corsair. In addition, company engineers were working on the UAC-funded VS-300 helicopter project.

The VS-44A "Flying Aces"

In December 1939, American Export contracted with Vought-Sikorsky for the design and engineering of a commercial version of the XPBS-1 flying boat, with an option to purchase three aircraft.

In July 1940, when AEA received permission from the Civil Aeronautics Board to commence trans-Atlantic passenger service, it immediately exercised the option on the flying boats. A contract was quickly executed between Sikorsky and AEA wherein three VS-44A aircraft would be built for the total price of $2,100,000.

Design work on the VS-44As began in

February 1940, and construction of the three aircraft began almost immediately. They were destined to be christened *Excalibur, Excambian,* and *Exeter* in honor of the three front-line cargo vessels owned by American Export's Shipping division.[8] Just as the cargo vessels were known throughout the company as the "Aces," the three aircraft became affectionately known as the "Flying Aces."

Officially, the three aircraft were designated:[9]

No. 1 *Excalibur*	Reg. No. NC41880
	Mfg. No. 4402
No. 2 *Excambian*	Reg. No. NC41881
	Mfg. No. 4403
No. 3 *Exeter*	Reg. No. NC41882
	Mfg. No. 4404

Extracted from company records, the following is a brief summary of the modifications required to convert the military XPBS design to the commercial VS-44A:

Plywood mock-up of the VS-44A in March 1940. (IISHA)

VS-44A Dimensions

Wing span	124 ft.
Height over Nacelles (Deck line level)	16 ft. 7 in.
Height over tail (Deck line level)	27 ft. 7¼ in.
Height over propellers on beaching gear	24 ft. 0 in.
Height over keel to sling in high position	21 ft. 3 in.
Length overall	79 ft. 3 in.
Dihedral center section	2 degrees
Dihedral	5 degrees
Chord-center section-maximum	20 ft. ½ in.
Incidence of Wing	5 degrees
Span of Tail	31 ft. 0 in.
Adjustment of Stabilizer	Fixed
Gross displacement of Hull	268,800 lbs.
Length of Hull	79 ft. 3 in.
Beam of Hull	10 ft. 0 in.
Height of Hull to Deck Line	11 ft. 11 in.
Diameter of propellers (3-blade)	12 ft. 6 in.
Tread of wing tip floats	78 ft. 0 in.
Maximum Draft	3 ft. 8 in.
Maximum draft with beaching gear attached	6 ft. 2 in.

Right: The Flight Engineer's station on *Excalibur* as delivered by the Sikorsky Company. (IISHA)

Below: Detail of the Flight Engineer's switch panel. (IISHA)

Above: Detail of the pilot's overhead control box. (IISHA)

The pilot's instrument panel, VS-44A. (IISHA)

Bow
A complete redesign was required due to the new rounded nose, and elimination of bomb sight doors and turret.

Bulkheads
Total number of bulkheads modified = 6
New bulkheads required = 3

Frames
Total number of frames modified = 31
New frames required = 11

Bottom Framing
Only minor changes were made in this area, but considerable drafting was necessary due to frame and bulkhead changes.

Side and Deck Framing and Skin
A major drafting effort was required due to changes in the locations of hatches, doors, windows, etc.

Interior
A complete redesign of the interior was required including the following:
(1) Flight Deck
(2) New Floors
(3) Soundproofing
(4) Heating and Ventilation
(5) Galley, Toilet, rooms, etc.

Beaching Gear
A complete redesign of the front and rear beaching gear was required.

Tail
The horizontal surfaces were enlarged and 10 degrees of dihedral added.

Controls
A complete re-routing of tail and aileron controls in the hull was required.

Wing Outer Panel
(1) Modifications to the leading edge ribs and skin for deicers was required
(2) Streamlining of the wing-tip float fittings was necessary to improve aerodynamics.

Wing Center Section
(1) It was necessary to remove the entire bomb and torpedo supporting structure and the lower surface bomb doors.
(2) It was also necessary to redesign the wing in the bomb compartment region and make provision for baggage compartments and doors on the upper surfaces.
(3) The leading edges were modified to accommodate deicers.

(4) Extra fuel and oil capacity was added.

Power Plant
(1) The nacelles and engine mount structures were reinforced for the larger engines.
(2) A CAA-required firewall was added.
Engine controls were revised due to the hull rearrangement.

Stress Analysis
A new stress analysis was required for the hull, wing, tail, beaching gear, controls, engine mount, and nacelles.

The hull was designed with two steps to facilitate take-offs.

Higher strength aluminum alloys and better knowledge of structures and loads enabled Sikorsky engineers to design this aircraft with a remarkable empty weight which was only 50% of the initial design gross weight of the aircraft.

An American Export Airlines advertising brochure aimed at potential customers claimed that "The VS-44s offered less vibration, less cabin noise, full-length beds, a snack bar, and many other conveniences which insured the utmost in comfort for the passenger."

A Sikorsky advertising brochure

ENGINE #1 IS PORT SIDE OUT BOARD

VS-4709 VS-44-A PILOT'S SWITCH BOX 3-20-42

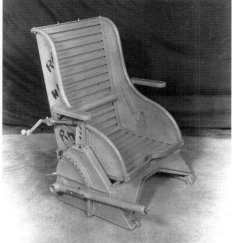

A close up detail of the pilot's chair, VS-44A. (IISHA)

VS-44-A RADIO OPERATOR'S PANEL 4-21-42

Above: Close-up of the pilot's switch box. (IISHA)

Right: Radio Operator's Panel on #41880 Excalibur. (IISHA)

released at the same time related more to the technical attributes of the aircraft by stating "The VS-44A is a high wing monoplane flying boat designed for non-stop trans-Atlantic operations carrying a full load of passengers, crew, and mail. The airplane has a top speed of 235 miles an hour, which is believed to be substantially faster than that of any other large flying boat now in existence. Its maximum non-stop range, under special fuel and load conditions, is in excess of 6,000 miles.

"It can be equipped to carry 32 passengers on daytime flights and under these conditions has a range of 3,000 miles at a cruising speed of 200 miles per hour. For non-stop trans-Atlantic service, it is equipped with sleeping accommodations for 16 passengers in addition to the crew.

"With a wing span of 124 feet and an overall length of nearly 80 feet, the VS-44A has a normal gross weight of 57,500 pounds."

"The wing is a full-cantilever type consisting of three assemblies—one center section and two outboard panels. The center section is a two beam arrangement with trussed ribs and metal covering from the leading edge. Flaps and ailerons are covered in fire proof fabric. The center section has three separate built-in compartments to carry approximately 4,000 gallons of fuel. Between the center tank and the outboard tanks, space is provided for baggage, mail, or additional fuel tanks. Two flares are housed in the center section. Nacelles for the four engines fair into the leading

edge of the center section. The nacelles are of the NACA type. The outer wing panels are of the same type structurally as the center section. They are readily detachable from the center section by removing four hinge pins. Deicer boots cover the entire leading edge of the wing, fin, and stabilizer.

"Other noteworthy design features of the VS-44A wing include a length/beam ratio of 7.6:1 as compared to the ratio on the S-42 of 6.9:1. The VS-44A has an increased wing loading of 35.5 lbs./sq. ft. The wing has an aspect ratio of 9.22:1 and a taper ratio of 0.25. Additionally, the wing has a ¼ chord sweepback angle of 6 degrees in 30 feet. Tail surfaces are full cantilever construction, structurally similar to the wing. The fin and elevator are metal covered, while the rudder and

Right: VS-44A Fuel controls. (IISHA)

VS-4920   VS-44-A FUEL CONTROL VALVES BULKHEAD 261
4-21-42

elevators are fabric covered. The hull is of aluminum alloy semi-monocoque construction, with six watertight bulkheads which enable the isolation of any one of the compartments. Hinged sections of the floor comprise the watertight doors for these bulkheads. Should the bottom strike a floating object and flood a compartment, the water would rise approximately only halfway up in the cabin. For this reason, the floor doors extend only a short distance above this flood water line, and it is thus impossible for personnel to be isolated in a partially flooded compartment. The hull is so designed that any two compartments can be so flooded and the ship remain on a reasonably level keel. Exposed riveting on the outside of the hull is both flush and modified brazier head type. Flush riveting is used on the deck plating from the bow to a broken line passing through station 123 at the chine and station 250 at the wing juncture, and on the bottom plating from the bow to the main step and from station 526 to the rear step. Brazier head rivets were utilized everywhere else.

"The hull has six openings besides its main entrance door; a mooring hatch in the bow, a bow cargo hatch, two pilot hatches, a wing hatch just aft of the rear spar, and a stern cargo hatch in the extreme rear end of the hull. Locking devices operable from either inside the hull or out are standard on all hatches.

"The forward end of the hull houses the crew, and includes sleeping accommodations for five crew members, the galley, mooring equipment, baggage storage space, and two men's rooms. The center portion of the hull is given over to passenger accommodations. The stern portion is arranged for a ladies' powder room, baggage storage, and accommodations for a stewardess. An interphone system enables the pilot to communicate with a member of the crew, whether he is at the bow, at the mooring hatch, or at the extreme stern of the hull.

"The passenger cabin is equipped with individual seats 40 inches wide and having adjustable cushions; full view windows, individual reading lights, and a complete heating and ventilation system. The seats are quickly convertible into comfortable beds. Upper berths are equipped with windows and reading lights. The men's lavatories and ladies' powder room are equipped with hot and cold water and complete toilet facilities. Provisions are made to accommodate three ten-person inflatable life rafts. As passenger comfort is a primary concern,

the aircraft specification allows for 400 lbs. of soundproofing and fabric panels to keep the sound levels below the following limits:

Control deck   85 decibels
Crew Quarters   82 decibels
Lavatories   80 decibels
Passenger compartments 72 decibels

"The heating and ventilation systems are built to maintain a temperature of 68 degrees Fahrenheit, even though the outside temperature may be as low as 20 degrees below zero. The source of heat is a "stove" built around the exhaust stacks of the two inboard engines. Temperature is automatically controlled by a thermostat located in one of the passenger compartments. An automatic air sampling system protects personnel from carbon monoxide by closing the hot air ducts and opening the cold air ducts. In addition, the flight engineer is warned of carbon monoxide presence by a red light system.

"An air scavenger system, with ducts

opening into each compartment, is built into the ceiling. There is also a separate fresh air system with individual valves controlled by the passenger. A complete galley is located just below the flight deck. Included in its equipment is a 7½ cubic foot refrigerator, a sink with hot and cold water, an electric stove with two hot plates, an oven, and ample storage space for dishes, pot and pans, and food.

"Equivalent to the bridge on a surface vessel, the flight deck houses the operating crew of the airplane. The operating crew consists of a pilot and co-pilot, a navigator, a radio operator, and a flight engineer who controls the operation of the engines and also controls an independent powerplant generating electricity for the many electrically operated units in the airplane.

"The power plant consists of four Pratt & Whitney Wasp air-cooled engines, developing a total of 4,800 horsepower for take-off. The engines are

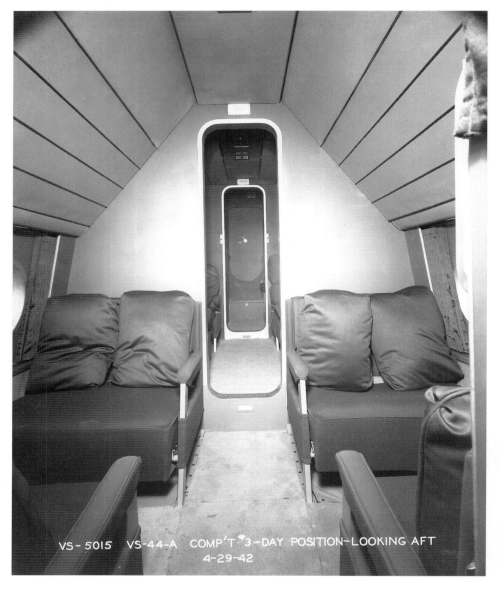

VS-5015   VS-44-A   COMP'T #3—DAY POSITION—LOOKING AFT
4-29-42

VS-5010 VS-44-A ICE BOX & SINK IN GALLEY   4-29-42

Above: View of galley sink and ice box. Stove is at left in photo. (IISHA)

Left: VS-44A Compartment #3 in daytime configuration. (IISHA)

Below: Close-up of galley sink, VS-44A. (IISHA)

fitted with three-bladed Hamilton Standard hydromatic quick feathering propellers with blades of the new laminar-flow section. The engine nacelles are of steel tubing, with aluminum alloy cowling. They are divided into three compartments by two firewalls. The forward compartment houses the engine cylinders, the center compartment contains the carburetor, magnetos, fuel pump, generator, and other accessories, while the oil tank is located in the after compartment. The engines are shock mounted in synthetic rubber. Quick disconnecting electrical plugs and pin-jointed engine mounts enable four mechanics to change an entire engine assembly in four hours. The fire control system is of the latest type, not yet installed in any other aircraft. A system of warning lights informs the flight engineer of the location of a fire and enables him to direct the extinguishing CO_2 to the spot through 22 outlets in each engine nacelle. The exhaust system of the engine incorporates the latest type of

ball and socket universal joints at each cylinder exhaust stack, enabling the maintenance crew to service a part of the engine without complete disassembly.

"The fuel system is so designed that fuel can be directed from any one of three tanks to the engines. Water which may have collected in the fuel tanks can be drained in flight, and fuel strainers can be cleaned in flight. In the event of failure of an engine-driven fuel pump, two electrically-driven pumps are standing by. Provision for dumping fuel, if necessary, includes a special scavenger system to clean out the dumping line of all traces of fuel.

"The aircraft is fitted with a detachable beaching gear which consists of three units—two located forward on the hull and one just aft of the second step. It is so designed that the airplane may be moved in any direction within the smallest possible space."

One of the first problems the engineers had to resolve on the VS-44A program was the undesirable stall char-

acteristics during landing which had been noted on the XPBS-1 program. Hull model tests conducted by the Navy and observations of the actual XPBS-1 indicated that during take-off and landings spray and even green water impinged on the propellers and particularly on the tail surfaces. More extensive hull model testing was conducted during 1940, both in the Sikorsky wind tunnel and in the Housatonic River. As a result of those tests, it was determined that modifying the XPBS-1 tail design by slightly increasing the size of both the vertical and horizontal surfaces and by adding a 10 degree dihedral to the horizontal surfaces, for the most part the aforementioned stall problems could be resolved. The 10 degree dihedral would raise the outboard tips of the tail by three feet. Further hull tests of the "V" tail indicated that, as predicted, this modification significantly improved the stall problem.

Additional wind-tunnel testing indicated that a vortex originating at the

VS-44A men's lavatory. (IISHA)

Detail of construction of upper berth frames of VS-44A. (IISHA)

View of crew bunks in galley-VS-44A. Hatch to Bow Compartment is at right. (IISHA)

intersection of the wing and hull was the cause of the fishtailing on both the XPBS-1 and the S-43. This condition was alleviated on the VS-44A by installing a vortex generator above the chine forward from the rear step. This generator was essentially an aluminum strip 30 inches long and 5 inches wide.

The first of the three VS-44As was completed on December 30, 1941. The aircraft was christened *Excalibur* at a ceremony at the Sikorsky plant on January 17, 1942. Mrs. Henry A. Wallace, wife of the Vice President of the United States, had the honor of launching the magnificent ship. Following tradition, Mrs. Wallace swung a bottle of champagne squarely against the bow of the aircraft. However, it bounced off undamaged. The aircraft's shock absorbent cork bow insert worked better than expected and the champagne bottle would not break. With a determined Mrs. Wallace standing by, a piece of steel tubing was quickly held against the bow. Mrs. Wallace repeated the act, neatly hitting the steel target and splashing champagne over the

aircraft and many of the invited dignitaries. Soon after, Mr. Sikorsky addressed the crowd. In a short and humble speech, he said "I had sincere expectation of great performance from the ship, which was built with one idea; to travel the greatest distance at the highest possible speed with maximum loads." AEA took delivery of the *Excalibur* immediately following the ceremony. The plane's first flight took place the following day.

The crew onboard the *Excalibur* for the first flight consisted of Capt. Charles Blair, AEA Chief Pilot, AEA Capt. Richard Mitchell, AEA Co-Pilot, Vought-Sikorsky employee Dean Taylor, Flight Mechanic, Vought-Sikorsky employee Donald Scott, Assistant flight mechanic, and AEA employee Michael Doyle, Ground Engineer.[10]

Perhaps Blair's account best reveals the excitement surrounding the flight when he stated: "I was only supposed to do high speed taxi on the step the first couple times out, but the aircraft just did not want to stay on the water." After a long "crow-hop," Blair eased the lively ship smoothly back on the water, then coaxed her airborne once again. She handled perfectly. Before landing, he flew *Excalibur* low over the plant "to show the workers what a magnificent aircraft they had created."

Following flight tests at the Sikorsky factory, *Excalibur* was flown to the Naval Air Station at Jacksonville, Florida, for

two additional months of testing which included CAA certification testing. During this phase of testing, conducted on the St. John's River in Jacksonville, several key take-off and landing parameters were established for the aircraft. At a gross weight of 57,500 lbs., the VS-44A could take-off and clear a 50-foot object with critical propeller windmilling in 6,250 ft. With all engines operating, and at 40,000 lbs., the aircraft could clear the same 50-foot obstacle in 5,000 feet. At gross weight, the distance required for a rejected take-off was only 8,000 feet.

In addition, Blair and two CAA pilots established that the take-off and landing characteristics were very satisfactory in both 2-foot and 4-foot wave conditions. The aircraft demonstrated a power-off lift coefficient of 1.8 with 0 degree flaps and 2.0 with 35 degree flaps. An impressive lift coefficient of 2.7 with 20 degree flaps was achieved at 70% power. This enabled the aircraft to achieve a landing speed of only 72.2 mph. The speed required to taxi on step was established at 55 mph. The aircraft was capable of accelerating to the take-off speed of 90 knots in 40 seconds. Also, while at Jacksonville, both static and dynamic testing was utilized to substantiate the aforementioned modifications to the tail.

In February 1942, as testing was nearing completion, Charles Blair summed up the aircraft's characteristics in a letter to American Airline (and future AEA) Captain Fran Wallace, that "The VS-44A

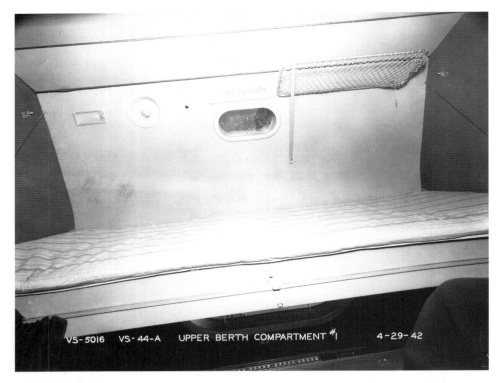

VS-5016   VS-44-A   UPPER BERTH COMPARTMENT #1   4-29-42

VS-5017   VS-44-A   LOWER BERTH COMPARTMENT #1   4-29-42

Top: Compartment #1 Upper Berth in down position with mattress, VS-44A. (IISHA)

Above: Compartment #1 Seats converted to lower berths, VS-44A. (IISHA)

is one excellent aircraft!" In April 1942, *Excalibur* was repainted in Navy camouflage consisting of sky blue on the top surfaces and light gray on the underside of the aircraft.

Excambian was delivered to AEA May 4, 1942, painted at the factory in Navy camouflage.

Excalibur made her first trans-Atlantic crossing on May 26, 1942, and began regular weekly round-trip service to Foynes, Ireland, on June 20.

Exeter was delivered June 23, 1942, also painted in camouflage. Like *Excalibur*, both *Excambian* and *Exeter* were placed in wartime service transporting passengers between Europe and America.

Flying the Atlantic

On June 22, 1942, *Excalibur,* commanded by Capt. Blair, left Foynes for her first official passenger-carrying flight westward to the United States. Blair was scheduled to make a fuel stop at Botwood, Newfoundland, some 14 hours into the flight. Blair recounted the flight: "We set out from Foynes on this longest of summer days, without any special ambition.

"There were 16 passengers in the passenger cabin—two in each compartment. They included Britain's famed combat admiral and commander of the Mediterranean Fleet, Sir Andrew Cunningham. The Admiral was on his way to Washington to help arrange the invasion of North Africa. We also had Dorothy Bohanna, R.N., the first airline stewardess to fly across any ocean.

"We dead reckoned below the clouds until past mid-ocean, using only the airspeed, compass, and drift flares for guidance. But dead reckoning wasn't enough. So, though it cost some fuel, we struggled up to 8,000 feet, where a sky full of stars informed us exactly where we were.

"Our weather messages told us that both Botwood and Shediac would be fogged in. Halifax, Nova Scotia was the final resort.

"These speed lines show we've made 30 miles in the last 30 minutes. That means we've got 80 knots on the nose. Our groundspeed is less than half our airspeed," said the navigator, Harry Lamont. "That's 4/10 of a mile per gallon and we've got 1,100 gallons."

"At that rate," I reckoned, "we'll have dry tanks a hundred miles short of Halifax."

"Glad this is a boat'" said the co-pilot, Captain Bob Hixson.

"We left the upper air and plunged through the clouds to find a better wind. The cloud ceiling was close to the surface of the ocean, but we still had to fly high enough to elude the masts of small shipping that might suddenly appear in front of us. At least the big icebergs were behind us."

"'Land-Ho,' said Bob Hixson. At this first landfall we had as many gallons of fuel left as there were miles to New York. I was beginning to toy with the idea of flying non-stop.

"'Mike,' I said, 'Recheck all those fuel records since Foynes—double check 'em. Maybe we'll go all the way.'

"Yes Sir! I'll have it in a couple of minutes." The co-pilot added his approval "Now you're talking, Charlie."

"We now had a few more gallons of fuel than there were miles to New York.

We had a reserve of a hundred gallons with a glass tube gauge. When the fuel started descending in that glass tube, we would fly one hundred miles more and that would be all.

"I let Halifax slip by as if it wasn't there. There were various small harbors and a couple of big ones ahead, and I could even land safely in the open ocean without any damage except to the ego.

"But New York's Whitesone Bridge was beneath us as we switched to the tank with the last hundred gallons of fuel. Our press agents had been boasting for years about flying non-stop across the Atlantic. Now, this was about to happen—the first time for an airliner with passengers and mail.

"The fuel tank showed 95 gallons when we looped the mooring rope around the bow post. This was enough, though it was only a fraction of the fuel we had on the river Shannon 25 hours and 40 minutes earlier."

"Remarkable voyage," said Admiral Cunningham of the British Navy as he disembarked, and the glint in his eye suggested he wasn't being loose with his praise.[11]

Also in Blair's *Red Ball In The Sky,* the author disclosed the fact that the normal westbound routes from Foynes to Botwood to New York or simply from Foynes to New York were unsuitable during the winter months. Strong westerly winds made the westbound crossing nearly impossible from mid-October through spring, even though the Sikorsky possessed the longest range of any flying boat operating in the Atlantic, including the Boeing 314. Blair stated that, "This winter journey via the South Atlantic covered almost triple the distance of the direct New York to

VS-4252   VS-44-A   ENGINE COWLING   1-24-42

VS-26SL VS-44-A ENGINE WITH MOUNT AND NACELLE - RIGHT OUTBOARD

Top Right: A close up of the starboard inboard engine (#3) nacelle and cowl on a VS-44A (IISHA)

Above Right: VS-44A starboard outboard engine (#4). (IISHA)

Right: VS-44A Port outboard engine (#1) showing close-up of mechanics platform (original caption on picture in error). (IISHA)

VS-4975 VS-44-A   WING TIP FLOAT - R.H.   4-25-42

VS-4979 VS-44-A BEACHING GEAR SHOWING HAND BRAKE

VS-44A Main Beaching Gear showing hand brake lever. (IISHA)

Above Left: VS-44A Starboard Wing Tip Float. (IISHA)

VS-4173 VS-44-A NO. II ENGINE - LEFT SIDE LOOKING INBOARD SHOWING MAGNETO BLAST TUBE 1-10-42

VS-44A port inboard engine (#2) showing magneto blast tube. (IISHA)

VS-3527 VS-44-A NACELLE - LOOKING FOWARD - R.H. 10-23-41

Close-up of VS-44A nacelle. (IISHA)

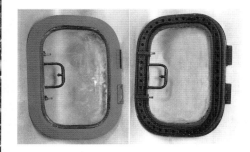

Interior and exterior views of VS-44A pilot's escape hatch. (IISHA)

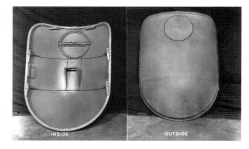

INSIDE   OUTSIDE

Interior and exterior views of tail escape hatch. (IISHA)

REAR

FRONT

VS-3659 VS-44-A INTERIOR WINDOW TRIM 10-28-41

Detail of interior window trim- VS-44A. (IISHA)

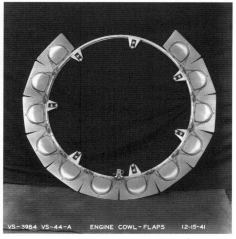

VS-3984 VS-44-A   ENGINE COWL-FLAPS   12-15-41

VS-44A engine cowl flaps. (IISHA)

Foynes. The first leg was southward 3,000 miles nonstop across the Bay Of Biscay to Portugal, thence along the coast of French Morocco, Spain's Rio Del Oro, and France's Senegal to Bathurst, British Gambia. From Bathurst, the route took a right-angle turn, heading west across the South Atlantic. It was 20 hours to Port of Spain, Trinidad, or a little less air time to Belem on the Amazon when the passenger load proved too heavy for the longer trip."

VS-4229 VS-44-A   CHRISTENING OF FLYING BOAT "EXCALIBUR"   1-17-42

VS-44A #NX 41880 outside the S-44 hangar at the Sikorsky plant two weeks before its first flight. (IISHA)

The formal christening of VS-44A *Excalibur* at the Sikorsky plant, January 17, 1942. (IISHA)

Above Right: The wife of Henry Wallace, Vice-president of the United States prepares to properly christen the *Excalibur*. Note the metal tubing taped to the cork bow. Repeated attempts to break the champagne bottle on the soft cork necessitated this temporary modification. (IISHA)

Another view of the christening of *Excalibur* with the Sikorsky seaplane ramp in the background. (IISHA)

During the first flight of *Excalibur*, Capt. Charles Blair made a low pass over the Sikorsky plant to show the aircraft off for those that designed and built it. (IISHA)

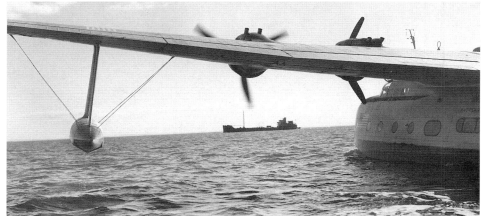

Right: The VS-44A water taxiing at the mouth of the Housatonic River. (IISHA)

Below: The *Excalibur* undergoing taxiing tests in the Housatonic River. IISHA)

Flying Tigers Return From China To Fly The "Flying Aces"

During 1942 several of the original American Volunteer Group (AVG) "Flying Tigers," who served under Claire Chenault in China, returned to the U.S., their military active duty requirement fulfilled, and went to work for American Export. This was an ideal situation for AEA as, being wartime, finding competent pilots was a difficult chore. The AVG pilots that came to work for AEA included: Peter Wright, Robert Layher, H. Geselbracht, G. Burgard, J. Hennesst, J. Cross, W. Fish, Robert Neale, R. Keeton, E. Beville, and F. Ricketts.

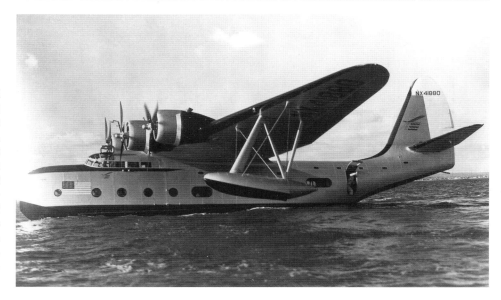

AEA Captain Fran Wallace. (Capt. Fran Wallace)

A close up of *Excalibur* early in 1942. (IISHA)

American Export Flight Engineer and restoration volunteer, John Liddell. (John Liddell)

Mechanic Harry Hleva carefully inspecting the beaching gear of the VS-44A in 1942. Harry would play a very important part in the history of the VS-44As. (IISHA)

AVG veteran and former VS-44 pilot Bob Layher shared some of his favorite memories of the VS-44A "Flying Aces":

In one eastward crossing Layher was alternating between the co-pilot's position and the navigator's position. Charles Blair was the pilot. Approximately one hour out of Laguardia the weather turned bad. They encountered a very thick cloud layer and visibility was basically zero.

Because of the clouds, the normal celestial navigation techniques were impossible. It was also impossible to even run the less accurate "sun line." At 1½ hours beyond their ETA at Foynes, the crew became concerned that they might be over the continent. For obvious reasons, in 1943 that could be a very dangerous place, especially for an unarmed transport plane. Blair, on the other hand, was convinced that they were not in trouble and in fact were near their proper destination. He chose not to turn back to the west as his crew urged but rather engaged in a circling maneuver until, after an extended length of time, he spotted a small break in the undercast. He spiraled the aircraft down through the break until they spotted water. They landed and found themselves in a lagoon framed on three sides with Irish countryside. Everyone dove for the maps. As it turned out, they were about 50 miles down the coast from Foynes and they were able to continue on with no further incident.

Another time, while nearly midway across the Atlantic, Layher thought that his time was up. Jim Craig was the pilot on this trip and Layher was serving as co-

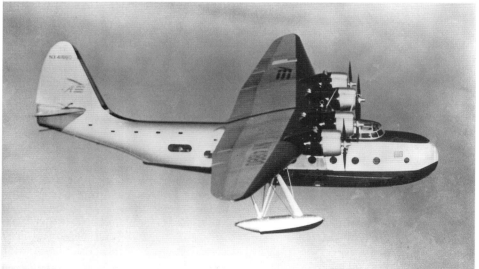

Left: The first pilot of the VS-44As and the last operational owner of a VS-44A, Captain Charles Blair. Charles F. Blair was born on July 19, 1909 in Buffalo, New York. His 45,00 hours flying career spanned civilian, naval, and air force. He commanded over 1,575 Atlantic Crossings and received many distinctions, including the Thurlow Award for navigation, The Harmon International Aviation Award for the World's outstanding aviator, and the Gold Medal of The Norwegian Aero Club.

On June 22, 1942, in the Sikorsky VS-44A flying boat, he flew the first non-stop flight with passengers and mail, from Foynes, Ireland to New York.

In 1944, again in the Sikorsky VS-44A, he piloted five consecutive fastest transatlantic crossings by sea plane.

On January 31, 1951, he recorded the fastest solo trans-Atlantic flight in his P-51 Mustang in 7 hours 48 minutes.

In May 1951, he made the first solo single-engined trans-Arctic flight using his "prepackaged" long range Celestial Navigation Technique for high speed aircraft. In 1952 he retired from the Navy to accept a commission with the USAF. He led and commanded the first flight of F-84F jet fighters non-stop across the Atlantic on April 18, 1956, with one in-flight refueling. On this flight he tested the two-star celestial navigation system which he had conceived. On August 7th 1959 he led and commanded the first jet fighters non-stop across the Arctic Ocean and North Pole (USAF operation "Julius Caesar") , which included three in-flight refuelings. He was awarded the DFC and commissioned Brigadier General. He died in a fatal aircraft accident on 2 September 1978. (Photo and biography courtesy of Maureen O'Hara Blair.)

Excalibur on an early test flight. (IISHA)

Right: Robert Quinn, Operations Manager, American Export Airlines. Bob was uniquely qualified for this position. In addition to being a licensed pilot, he was also an A&E, a CAA licensed meteorologist, and had a B.S. in Aviation Operations. Bob is currently on the Board of Directors of the Glenn L. Martin Aviation Museum in Baltimore, Md. (Robert Quinn)

American Export Airlines International Stewardesses. Chief Stewardess Dorothy Bohanna, the first stewardess to fly across an ocean is second from right. (Capt. Fran Wallace)

AEA stewardesses Tony Serrell and Genevieve St. Mary pose on horizontal stabilizer of *Excalibur*.

AEA International stewardesses strike an appealing pose next to the VS-44A. (Capt. Fran Wallace)

pilot. Everything seemed normal so Layher ventured down to the galley to get some coffee. In the short time that he was there, the aircraft's wings iced up and at an altitude of 10,000 feet the aircraft went into a stall. It took everything the pilot could do to bring the aircraft under control. They were less

Two views of the wing of VS-44A aircraft #41880. Parts of the *Excalibur* have been retained for display at the Botwood Heritage Museum. (IISHA)

AEA Stewardesses Margaret Siegfried, Kathryn Friedrich, Genevieve St. Mary, and Adele Jenkins. American Export Airlines was the first over-ocean air service in the world to place stewardesses aboard aircraft as regular crew members. All American Export Stewardesses were registered nurses and every one of them had previous experience as airline stewardesses. (Bob Quinn)

than a 1,000 feet above the water when he finally recovered from the stall.

Bob also has fond memories of a trip to North Africa to deliver delegates to a summit conference at Casablanca being held by President Roosevelt and Winston Churchill. The trip over was uneventful but in the exceedingly hot, very calm weather of North Africa, when it came time to leave, it took the VS-44A five or six attempts to finally get airborne. During each of the aborted take-off attempts, Roosevelt, Churchill, and all of the delegates stood along the rail of the U.S. Navy heavy cruiser that they were

Actor Humphrey Bogart and his wife Mayo Methot on their return to New York after an extensive USO tour. The Bogarts were frequent flyers on the VS-44As. (Bob Quinn)

meeting on and cheered encouragement to the big flying boat. Needless to say, this distinguished audience added just a little pressure to the already very busy flight crew.

Excalibur Crashes

On October 3, 1942, *Excalibur* (N41880) crashed on take-off from Botwood, Newfoundland. Botwood was the most utilized intermediate stop on the crossing from the U.S. to Ireland. Reportedly, the aircraft porpoised two or three times, finally gained about ten feet of altitude, settled back on the water, then took off again with a nose high attitude of about 30 degrees. After attaining an altitude of approximately 35 feet, the plane gradually nosed down and struck the water at an angle of about 35–45 degrees. The aircraft broke apart into several sections. Parts of the aircraft settled to the bottom of the Bay of Exploits but many parts

were recovered and are currently displayed at the Botwood Heritage Museum. At the time of the crash, there were 26 passengers and a crew of 11 onboard. Six passengers and five crew members perished. The official report lists inadvertently lowered wing flaps as the reason for the accident, but Captain Blair, who was then Chief Pilot for AEA had a much different theory.

In 1978, Capt. Blair related to Sikorsky Engineer Ralph Lightfoot that in his opinion this is not what happened at all. He said that he had been having difficulty with the pilot in that the pilot did not always follow proper operating procedures and tended to operate as he pleased. Blair stated that "The man was a big strong man who felt that he could override any effects of full flap position upon the aircraft and that he had intentionally set full flaps for take-off. Of course, once the airplane was airborne, it shot upward in the air and the aerodynamic effects of the flap suddenly swung the nose down to the water and the pilot didn't have enough space and time to recover the airplane. Consequently it crashed."

As a result of this accident, the flap position operations were later changed to a three way switch. Forward on the switch was for 10 degrees of flap. Straight up was neutral and full aft was for 20 degrees. It was now not possible to set the flaps in any other positon. Full aft on the original flap control was 35 degrees.

While the VS-44 was generally regarded as being well mannered, she did have some "special" traits. Regarding take-offs, Capt. Blair commented, "The VS-44 was a real submarine; it sat so low

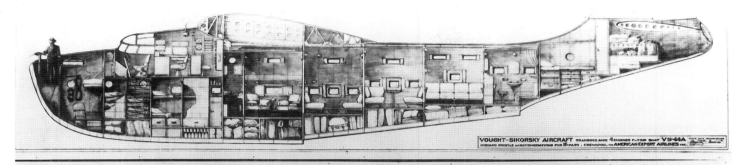

Above: VS-44A cutaway drawing, 1942.

Right: The *Excalibur* in flight. (IISHA)

that the water almost reached the windows when we carried a full load. When we applied take-off power, the spray of water nearly submerged the ship. Take-off was critical. On hot days, we would use up about two miles of water getting airborne."

AEA Stewardesses Were First To Cross An Ocean

No discussion of the "Flying Aces" and American Export Airlines would be complete without giving credit to the ladies whose job it was to be mother, sister, nurse, waitress, upstairs maid, companion, cheer leader, or extra hand at cards for the passengers aboard the *Excambian* or *Exeter* on transatlantic trips.

AEA was pioneering transatlantic passenger service and the AEA stewardesses were a very large contributor to the success of that initiative. The airline required that stewardesses on these flights be registered nurses with exceptional nursing experience, have at least 20 months of experience as a stewardess on domestic flights, and pass a very rigorous physical examination.

On June 30, 1942, Dorothy "Dottie" Bohanna made the first transatlantic trip as a stewardess on a commercial airline. By late 1942, a total of eight ladies were assigned as "stews" on these flights. On a typical flight which could last as long as 25 hours, the stewardesses were usually on their feet almost the entire time; hence they used to joke that they "walked to Europe." They had very little opportunity to cat-nap during the trip and as a rule could expect only 2–3 hours

of sleep. Because of the duration of the flight, passengers were served three complete meals. These meals were prepared from scratch and served, in the style of the finest restaurants, to the 26 passengers on board.[12] The nursing background that these ladies possessed was no luxury on transatlantic flights. Each airplane had an adequately stocked medical chest. Passengers had to undergo a physical examination before embarking on a crossing and stewardesses were responsible to render any medical help required during the flight. Stewardesses on domestic flights did not need this very special skill simply because on over-land trips, if a medical emergency arose, the pilot could land at the nearest airport and summon medical help. Obviously, transatlantic pilots did not have that luxury! In addition, at the conclusion of transatlantic flights, the stewardess had the responsibility to certify to immigration authorities that the passengers had no health problems. This was quite a lot of responsibility for a position that started at $165 per month!

Naval Air Transport Service

The U.S. Navy requisitioned the two remaining "Flying Aces" in January 1943, but contracted to have AEA crews both fly and maintain the aircraft. That decision, coupled with leaving AEA markings on the aircraft, allowed the aircraft to operate out of neutral ports.

Also during this period, as the Navy realized the value of the large flying boats for trans-oceanic transport, they established a requirement for approximately 200 VS-44As to be designated the JR2S-1. Because the Vought plant was fully dedicated to building F4U-1 Corsairs and the new Sikorsky plant in Bridgeport was dedicated to the production of the R-4 helicopter, the Navy decided to have Sikorsky build (in the Stratford Plant) the first six aircraft and to award a contract to Nash-Kelvinator to build the remaining 194 under the designation JRK-1.[13] The primary difference between the JR2S-1 and the VS-44A was the addition of a large cargo door. Sikorsky began fabrication of the first six hulls under U.S. Navy contract #93164,

Excalibur repainted in Navy camouflage in April, 1942. (IISHA)

and Nash-Kelvinator began construction of a large facility to fabricate their share of the order. Unfortunately, the contract was canceled, and the six hulls were scrapped. At that time it was speculated that the contract was canceled because the Navy wanted to concentrate on producing the PBY Catalina. Another popular explanation was that the Navy was transitioning to an emphasis on land-based planes rather than seaplanes. In a short history of the VS-44As written by Hugh Wells sometime around 1948, he speculated that "The other big flying boat manufacturers complained that United Aircraft Corporation had more than their fair share of government contracts. The air industry knew that it was a matter of policy, not failure. The VS-44 design easily met all U.S. Navy requirements." However, the most plausible explanation for the contract cancellation was that the Consolidated Coronado, which six years earlier had bested the XPBS-1 in the production award competition for a large flying boat patrol/bomber, had proven unsatisfactory in that role. The Navy was reconfiguring them into cargo aircraft, and simply did not need the big Sikorskys.

The "Flying Aces" Go To War

During the war years, *Excambian* and *Exeter* continued to ferry passengers, cargo, and mail across the Atlantic to such ports of call as Foynes, Bathurst, British Gambia, and Port of Spain, Trinidad. Many VIPs, including Queen Wilhelmina, Eleanor Roosevelt, Humphry Bogart, Douglas Fairbanks Jr., Edward G. Robinson, and General Omar Bradley traveled on the VS-44As. It is known that many Jewish refugees were transported on the VS-44As to the United States during the war.

On December 31, 1944, American Export Airlines' contract with the U.S. Navy was terminated. In January 1945, *Excambian* and *Exeter* were repainted in original American Export Airlines colors and placed back in scheduled passenger service. In June of that same year, AEA merged with American Overseas Airlines which later became American Airlines. Shortly thereafter, the airline discontinued use of the remaining two "Flying Aces" in favor of new land-based DC-4s. The number of landing strips built during the war, both in this country and overseas, made the landplane more practical and more important, more economical. The last trans-Atlantic flight of the VS-44 was on October 22, 1945, from Foynes to New York. That was also the last American Export flying boat and the last passenger-carrying flying boat to depart from Foynes. Captain Charles Blair was the pilot and before he departed, he circled the town and made a low pass down the length of the main street. In terms of the major air carriers, the era of the giant flying boat was over.[14] Late in 1945, the airlines retired both *Excambian* and *Exeter* and the Reconstruction Finance Corporation put them both up for sale.

In 1946, both aircraft were purchased

CAA Certification Granted

On July 14, 1943, the CAA granted the VS-44A an approved Type Certificate #752 (transport category) with the following limits:

Maximum take-off gross weight	59,534 lbs.
Airspeed limit	185 mph
Maximum landing weight	51,809 lbs.
Airspeed limit	211 mph
Fuel capacity	3,820 gal.
Maximum baggage weight	9,920 lbs.

Engines: Four Pratt & Whitney Twin Wasp S1C3-G
Engine limits: Max except take-off: 1,050 hp/engine
Take off limit: 1,200 hp/engine (2 minutes)

Airspeed Limits (True):

Gross Weight	Level Flight or Climb	Glide or Dive	Flaps Extended 20°
51,500lbs	211 mph	236 mph	144 mph
59,225lbs	185 mph	221 mph	144 mph

VS-5478   VS-44-A   VS-44-A 3/4 REAR LEFT HAND   6-15-42

Exeter, Number 41882, at the Sikorsky plant prior to delivery. (IISHA)

A photo which illustrates the volume of mail carried overseas to the troops by the VS-44As during World War Two.

from the War Assets Corporation by Huestis Wells and his New Orleans/Tampico airlines and were immediately placed into charter service. On July 13, 1946, *Excambian*, while under charter to Condor Airlines of Peru, made a non-stop flight from Lima, Peru, to New York City, making the trip in 19 hours 56 minutes and covering a distance of 3,300 miles. Hugh Wells was the Captain and pilot during that flight. It is known that Tampico subsequently experienced difficulties with its finance company, Seaboard Financial, and consequently both *Excambian* and *Exeter* were sold to Skyways International in April 1947. Charles Blair, who was employed by Pan Am at the time, was hired by Skyways to check-out their pilots in the VS-44.

In June of, 1947, Charles Blair leased *Excambian* from Skyways to ferry men and supplies from Minnesota to Iceland. He took delivery of the aircraft in Baltimore and had the company name "Associated Air Transport" painted on the fuselage. *Excambian* was renamed *Reykjavik* in deference to the capital of Iceland. During the lease, approximately 300 men were transported to assist the Icelandic government in the construction of a new hangar and terminal facility at the Keflavik Airport.

In *Red Ball In The Sky*, Charles Blair recalled the first of the flights in the *Excambian* to Iceland. "On the first flight out of Lake Minnetonka we carried an unusual group of passengers, of which a few looked as though they might have escaped from the local jail. It appeared that their fear of the air had forced some of these delicate souls to over-consumption of spirits.

"Getting these inebriated gentlemen into the unsteady shoreboat and then disembarking them onto the flying boat was a hazardous affair. But not a soul was lost, although some did get wet. We eventually got everyone safely aboard, with the difficult ones stowed in the rear passenger compartment where they were

Records Established by VS-44As Flown By AEA

While in service with American Export Airlines, the "Flying Aces" established several notable records.

The "Flying Aces" were the longest-range commercial aircraft in the service of any airline, and were the only aircraft that flew commercial schedules non-stop with a capacity payload across the North and South Atlantic on flights in excess of 3,100 miles.

1. Transatlantic Record, U.S.A.–Europe: 3,329 miles in 14 hours 17 minutes (non-stop).[15]
2. First to fly non-stop Foynes to New York, June 22, 1942: Captain Charles Blair, pilot, Captain Bob Hixson, co-pilot.
3. First non-stop flight New York–Lisbon: 3,383 miles in 20 hours 14 minutes, Captain Charles Blair.
4. First to fly non-stop Baltimore–Europe: 3,380 miles in 16 hours 2 minutes.
5. Fastest westbound Atlantic Flight time, Europe–U.S.A. (With refueling stop at Botwood, Newfoundland) in 17 hours 45 minutes.
6. Fastest non-stop flight between Europe–New York in 18 hours 5 minutes.
7. First to fly non-stop Bermuda-North Africa: 3,362 miles.
8. First to fly non-stop Bathurst (Africa)–Port of Spain (Trinidad): 3.109 miles.
9. First to fly non-stop Bathurst (Africa)–San Juan (Puerto Rico): 3,384 miles.
10. Fastest westbound crossing Foynes to Laguardia (With refueling stop at Botwood): 17 hours 57 minutes. Capt. Edward A. Stewart, pilot.
11. Fastest non-stop flight New York to Foynes: 14 hours 17 minutes. Capt. Charles Blair, pilot.

VS-44A #41882 *Exeter* at Floyd Bennett Field, date believed to be early 1947 shortly before aircraft was purchased by Skyways International. (IISHA)

VS-44A #41881 prior to restoration at Baltimore, 1948.

segregated from the sober passengers.

"But peace was short-lived. Soon after take-off my bedeviled and exasperated purser, Mario Borges, came scrambling up the ladder onto the flight deck to advise that pandemonium was running rampant in the rear compartment over the possession of some whiskey bottles which had been sneaked aboard.

"I decided the best course of action would be to climb to 12,000 feet, where the shortage of oxygen in the Sikorsky's unpressurized cabin would further addle the intellects of my unruly customers and render them reasonably docile. After reaching this altitude I proceeded aft to separate these passengers from the source of their problem. In the rear compartment I found a half dozen shirtless gorillas entwined among each other like a mass of snakes, wrestling listlessly for their beloved bottles."

"I had no difficulty digging the offending spirits out from under this pile of wriggling, sweating humanity. Whatever strength they once possessed had been reduced to total flabbiness by oxygen starvation. A stern warning even caused them to fasten their seat belts like obedient children."

Exeter Crashes

While *Excambian* was plying the northern-most flying routes under the command of Captain Blair, her sister ship *Exeter* was occupied in equatorial regions performing much more clandestine activities. On August 15, 1947, while ferrying arms and ammunition to Paraguayan rebels, *Exeter* crashed while attempting a night landing on the River Plata near Montevideo.

Cornelius J. "Ken" Dineen, VS-44 flight engineer and restoration volunteer, survived the crash of the *Exeter*. He recalls that "The Uruguayan President called Skyways and talked with us, saying that he would assist Skyways in forming a Uruguayan Airline if we could help assist the Paraguayan rebels. What they wanted us to do was to take a couple of loads from Montevideo to a point on the Parana River where Brazil, Paraguay, and Argentina come together. We were supposed to rendezvous with a gunboat on the river, unload, anchor for the night, and take-off and head back the next morning.

"We made one trip and all we carried was military supplies. As soon as we landed on the river, a gunboat came over. It was nothing as big as a destroyer. As we unloaded, which took all night, this gunboat stayed with us. The river was patrolled by Argentina and Paraguay, and the boat gave us protection. At sunrise, we cranked up the engines and got back to Montevideo.

"We were up most of the next night, loading the VS-44. The Uruguayan Navy brought out Brazilian arms and ammunition to go aboard. The following night, with so much goddamned stuff on board, we tried to take-off but could not

The VS-44A from starboard aft. (IISHA)

VS-44A *Excambian* in Navy camouflage with American Export markings. (IISHA)

even get up on step. We went back to the mooring, contacted the Uruguayans, and they had to take off a lot of stuff. We also had six Uruguayan doctors aboard, to help treat the rebels in Paraguay.

"That night, we did take-off carefully and flew up to the rendezvous point. We found that the gunboat we were supposed to meet had gone up the river in Asuncion. There was a battle going on there between the rebels and the government forces. We didn't dare land...or we probably would have ended up in an Argentina prison...or dead. So we headed back to Montevideo. It was dark...and I mean dark.

"The landplane base at Montevideo was a Pan-Am operation. We tried to contact them by radio to no avail. No one would answer to give us an altimeter setting. We couldn't see the water and had to just feel our way down.

"You've got to understand that landing a seaplane at night is like nothing else in this world! The maximum descent rate should be about 150 feet per minute. Normally you go in until you stall the airplane just before you touch the water. In those days, the old boats would land at about 110 miles per hour max. With no altimeter setting and no visual reference to the water, you don't know when to pull back for the stall. Lights don't make a bit of difference.

"Dick Granier was pilot, Andy Fisher was co-pilot, Billy Miller was radio operator, and Hoagy Jeertsen was purser. We also had a Uruguayan pilot, myself, and six doctors aboard.

"Dick got on the interphone. 'Ken, how do I use the landing lights?' I thought that he was kidding. I answered, 'Why don't you try aiming one light down and the other one out?'

"I was in my seat. The landing light on the left wing was pointed down. I looked out the window. Water seemed to be coming up awful fast. I turned in my seat...I had my safety belt on...and said, 'Hey'...and we hit!

"Sitting sideways, I had nothing in front of me. I went over and down so that my head was below the flight deck. I saw the bottom of the deck. Water was coming in all over the place. When I saw the water rushing in, I knew that the hull plates were gone. There were steps that went below the flight deck. I struggled up and got a look forward. I know that I cried out. I guess I was in shock.

"It was deadly quiet as the airplane started to go down. I sat in my seat and told myself not to get loose with everything surging around. I felt the engineer's panel go off over my head and saw all kinds of wires and cables swaying. Ballast for the wing came off and tipped up at the leading edge, with part of it hanging off. And then the whole wing began peeling off.

"The wing on the VS-44 is held on by four one-inch bolts. As we tipped, evidently the back bolts broke first, the wing pivoted on the front until it came loose, and it hung vertically with the four engines on it. With nearly empty gas tanks, a few cartons of grenades and rifles in the hold, the wing pulled down and ripped off.

"The hull, I believe, went down vertically (with the wing off) and then rolled on its left side. It was about 40 feet deep there. Thinking back, we weren't very far from where the German Battle-Cruiser *Graf Spee* was sunk.

"I released my belt, got tangled in wires and stuff, and began to pass out—almost like going to sleep. I came to, realized that I was under water...and I dived

Mr. Ken Dineen's Flight Engineer wings for Skyways International Airways. (Erwin Botsford)

Right: VS-44A #41881 prior to restoration at Baltimore, 1948.

VS-44A *Excambian* prior to launch in Baltimore after $240,000 restoration.

down and somehow rose to the surface. The bottom of the wing was next to me with the rear spar showing about four feet out of the water. I climbed and stepped on stringers and got up on the spar. A rubber bunk mattress wrapped in a blanket rose next to me. I got the rubber mattress and wrapped it around me. I had no life jacket on. My right leg was cut up and bleeding. I took my belt off and made a tourniquet out of it.

"A Uruguayan tug boat appeared and we were able to jump on the deck. They got us ashore where we were taken to a navy hospital. I had broken a big toe and cut my leg pretty badly. We found out that only three survived. Dick Granier, Andy Fisher, Billy Miller, Hoagy, and five Uruguayan doctors all had perished. I was in the hospital about two weeks."

"Charles Blair had personally checked out Dick Granier, the pilot, for day-time operations but unfortunately, the pilot had no night landing experience with the big flying boat. That, coupled with the fact that the aircraft was severely overloaded, spelled disaster."

In the 1980s, scuba divers located the submerged wreck of the *Exeter* and reported that the hull was reasonably intact with the left half of the wing broken off but lying nearby covered with river mud. The divers recovered a few of the rifles that never made it to the rebels.

Excambian Changes Hands—Again

After the Iceland lease, *Excambian*, now the only remaining "Flying Ace," continued in charter service until 1948, when she was impounded by the city of Baltimore for non-payment of storage fees. It is believed that the aircraft was purchased by a local minister for approximately $500. Sometime later, the minister sold the aircraft to the Aviation Exchange Corporation headed by Hugh Wells. *Excambian* was subsequently refurbished, at a total cost which exceeded $240,000.

Heustis Wells is significant in the history of the "Flying Aces." His name pops-up repeatedly as having been connected to the *Exeter* and *Excambian* beginning with the post American Export Airlines days right up to the days before Avalon Airboats acquired Excambian. As architect of this major refurbishment, he probably expended more money than any other owner to keep the aircraft flying. "Hugh" Wells was an aviator trained by the U.S. Army in the 1920s. He was a native of Worcester, Mass. His 1964 obituary claimed that he had nearly 1,000,000 flight hours. That is impossible, but what is certain is that Wells had a very colorful aviation career. He is credited with pioneering many new air routes in South America. As a pilot for Pan American Airways, he captained the first scheduled international flight by a U.S. carrier, carrying mail from Miami to Havana, Cuba. Ed Musick was his co-pilot on that flight. At one time, he was indicted by the U.S. government for allegedly attempting to sell military bombers to the government of Bolivia. It is unclear whether he was ever convicted.

Robert Scott, who worked on the flying boat from 1952–1955 while it was undergoing refurbishment at Baltimore, shared his remembrance of the refitting process. He wrote "A man named Hugh Wells bought the VS-44A from the city of Baltimore and was having work done on it at Harbor Field. Baltimore Aero Service, owned and operated by Turfield Miller, recovered all fabric surfaces, began wringing out all electrical wiring and cleaning the airplane to determine the extent of corrosion and repairs required as a result. Mr. Wells decided to lease space on the airport, hire a few full-time people, of which I was the only licensed mechanic. He used part-timers from Glenn L. Martin Co. and from the Air National Guard. The engines were overhauled by Airwork, Inc., in Millville, N.J. I don't remember if Airwork or someone else overhauled the props.

"A new generator control system was designed by a Martin engineer, and new wiring and control panels were installed by Martin electricians. Extensive structural repairs were required in the hull bottom and bottom of the center fuel tank. The three tanks were integral tanks and required cleaning and re-sealing.

"The fuselage was painted a light metallic blue, the hull bottom was red, and the upper surface of the wing was international orange. All fabric surfaces were silver. On the side of the hull, forward, was the word *Excalibur* in script about eight or ten feet long, in silver leaf, trimmed with red. I wish that I had taken some color pictures of it.

"A few local test flights were made and then the trip to Boston. I never did get to see the plane fly because I was always either flying it as co-pilot or acting as the flight engineer."

The Aviation Exchange Corporation intended to use the ship as a flying trading post, transporting goods to the natives of the Amazon River and returning with precious gems, alligator skins, and quinine. The plan proved to be economically unfeasible and the *Excambian* ended up being put in storage at Ancon Harbor, Peru.

Avalon Air Transport

Excambian remained in Peru until 1957. During the spring of that year, "Dick" Probert and Walt von Kleinsmid, co-owners of Avalon Air Transport obtained *Excambian*.

"Dick" Probert related the story of how *Excambian* joined the Avalon Air Transport fleet:

"Sometime in the spring of 1957, an insurance agent for Avalon Air Transport presented to Walt von Kleinsmid, my partner in AAT, and I a picture of a four-engine flying boat which appeared to be sitting on its beaching gear on some seaplane ramp. The picture, a color photo of N41881, was printed on a card the size of a normal business card. Walt and I showed some interest in this flying boat until the insurance agent quoted a price of $250,000. At that point, our interest evaporated. We couldn't even think about that amount of money. Several months went by when we received a phone call from a person in Baltimore who wished to talk to us about a four-engine flying boat. I still had the picture that the insurance agent presented to me and after describing to the gentleman calling what I observed in the picture, we both agreed that the airplane we had a picture of was, in fact, the airplane that he wanted to talk about. Rather than enter into a long-winded conversation over long distance phone lines, I planned to cut the conversation short by asking the gentleman what he had in mind as a price for this airplane. When he quoted me a price of $125,000 my adrenaline started running for I was sure we could handle that amount. Both Walt and I were definitely interested and as a result we made an agreement with this individual to come to Long Beach to formulate a purchase plan satisfactory to all concerned for our purchase of the flying boat.

"According to this gentleman, the airplane was presently in Ancon, Peru, a small town about 30 miles north of Lima. The previous owner, Aviation Exchange Corporation, was primarily operated by a former U.S. military pilot by the name of Hugh Wells. Wells had cleared a beach area at Ancon, which

when using about one half the normal air pressure in the beaching gear could be utilized as a seaplane ramp. This airplane would actually taxi over the wet sand into and out of the water with this very low pressure in the beaching gear tires. N41881 had been on this beach for 18 months. According to information furnished by Well's flight engineer, Henry Ruzakowsky, the engines had been run often enough to keep them in good operating condition.

"When I first observed this airplane, it was sitting on this ramp headed away from the water and with both the tail beaching gear wheels missing. The tail was supported by the stump of the tail beaching gear structure. This was on April 28, 1957. The gentleman with whom I was to make all future plans, the owner's representative, John Forrester, who, on this occasion, was with me, advised that the previous flight engineer felt the company still owed him some money and he had removed the tail wheels to be sure that the airplane did not leave until he had been paid.

"All this was occurring at the time when the non-scheduled air carriers in the USA were having a hey-day, which, among other things, included adding additional seats to the interiors of planes then in use on continental flights, mostly DC-3s and DC-4s. This was done by installing smaller, less comfortable, and lighter seats and by making the

aisles narrower to accommodate the extra seats. DC-3s which were equipped with 21 seats by the factory, were changed to 32 passenger seats. This could be accomplished legally by keeping the baggage weight and the weight of the fuel low. A DC-3 carries gas tanks that hold 808 gallons of fuel. I flew many trips with only 300 gallons in the tanks on take-off. That, of course, meant many stops for fuel.

"With this in mind, I visualized a seaplane that would hold 75 passengers. This, of course was predicated on the weights furnished me by the sellers. The difference between the empty weight and the gross take-off weight of the aircraft was such that with only 1,000 gallons of gasoline; and with 100 gallons of oil and a crew of four, 75 passengers would bring the take-off weight to 10,000 pounds under the allowable gross take off weight. At this point, I could see Avalon Air Transport giving the Catalina Steamer some real competition.

"This whole plan went out the window when I first stepped inside the airplane. I had had visions of a clean interior like that of a DC-3, DC-4, DC-6, or Constellation. This plane had bulkheads, a total of six, throughout the aircraft. These bulkheads are a must for large seaplanes. Besides adding considerably to the strength of the aircraft, they are a built-in safety requirement to prevent the aircraft from sinking. If any

Avalon Air Transport owner Dick Probert with wife and chief stewardess Nancy Ince Probert and VS-44A Captain Lloyd Burkhard. (Dick & Nancy Probert)

Avalon Air Transport owner Dick Probert scans the "water runway" for floating debris before take-off.

particular compartment flooded, it could be easily and quickly sealed and isolated. With these bulkheads in the aircraft, high density seating was impossible.

"The purchase agreement which Walt and I had entered into was in two parts; Part A and Part B. Part A required that I would go to Lima, at Avalon Air Transport's expense, inspect the aircraft, and if satisfied that we could use it, we would purchase it as is, where is. The agreed price was $125,000. Part B, which would become effective if, after my inspection, I felt we could not use it, required me to ferry the airplane to Long Beach, participate in the sale for 25% of the sales price, and take all of the expense of ferrying to Long Beach off the top of the sale price when sold.

"John Forrester, the owner's representative, was with me at the time I first inspected the airplane and I therefore had the unpleasant task of telling him that we could not use the airplane and would have to go to plan B. There were numerous items which I demanded be accomplished before I would fly the airplane. I advised John Forrester that I required a new annual inspection, required him to acquire the tail wheels and get them installed, and make arrangements with Henry Ruzakowsky, who was still in Lima, to engineer the flight to Long Beach. John and I agreed that I should go to Long Beach while he carried out my requests. I, Therefore, left Lima for Long Beach, arriving in Long Beach on May 3, 1957.

"On May 5th, John Forrester called from Lima to advise that all of my requests on the Sikorsky had been complied with. The following day, Monday, May 6th, I was on my way to Lima arriving there on May 8. These were the days before jet transports and such flights as I was making were made in DC-6s, which took a couple of days to do what the jets do today in a few hours. In addition, the flights to Lima from Los Angeles proceed via Rio de Janeiro.

"Upon arriving in Rio, I proceeded directly to Ancon where I found that the wheels and tires for the tail beaching gear were not on the gear. John Forrester,

The pilots instrument panel as configured by Avalon Air Transport. Note that the flight engineers panel was eliminated and the engine controls were relocated between the pilot's and co-pilot's seats. (Proberts)

Dick Probert uncaps the exhaust ports prior to spring start-up of *Excambian*. (Proberts)

who was at the airplane when I arrived, advised that he had made a deal with Ruzakowsky to bring the wheels to the airplane and install them. Ruzakowsky had also agreed to engineer the necessary test flights as well as the flight to Long Beach. I agreed to purchase airline tickets from Long Beach to Baltimore and from there to Lima for Ruzakowsky. Unfortunately for me, the Sikorsky had been manufactured with a flight engineer's panel and it was therefore impossible to fly the airplane without a flight engineer.

"With Ruzakowsky available, I hired him on May 10, 1957, we launched the aircraft, placed it on a mooring and removed the beaching gear. After checking the hull for leaks, checking the fuel and oil supply, we flew her. She flew remarkably well, was normally heavy on the controls, as were all large aircraft built without boosters on the controls, but was a definite pleasure to fly, until, 17 minutes into the flight when engines number two and four started to backfire,

run exceedingly rough, and within two minutes, quit altogether.

"Except for four single reading tachometers and four manifold pressure gauges, all of the instruments were on the flight engineer's panel. I would, at that time, preferred to have had in my view the cylinder temperature indicators, the oil temperature indicators, the oil pressure indicators, and the fuel pressure indicators. These were all on the flight engineer's panel, so I was at the mercy of the engineer to tell me what he thought had happened to make the two engines quit all at once. He advised that the cylinder temperature indicators went from operating temperature to zero between the time the engines started back-firing and when they quit. This situation left me in the air with an airplane that I was not familiar with, in a territory that I was not familiar with, and to make a landing in a landing area that I was not familiar with, and with two engines not functioning on a four-engine airplane!

"Unfortunately, the two engines which were not operating were number two, an inboard, and number four, an outboard. As long as the two engines which were not operating were not on one side of the airplane there was no problem in controlling it in the air. However, water handling was a different situation. After landing, which really presented no problem, taxiing to our mooring was a bit challenging. The inboard engine would not turn the airplane, so if a turn to the left was required, it was accomplished by making a 270 degree turn to the right. After much twisting and turning we had the airplane at the mooring.

"All of the foregoing was reported to John Forrester that evening upon our arrival at the country club of Lima where all concerned with this airplane were staying. John pleaded with me not to return to Long Beach, saying that he had two excellent mechanics who would correct the problem. John's two excellent mechanics were the chief of mainte-

VS-44A   January 11, 1958

To Igor Sikorsky
Best wishes from
Avalon Air Transport, Inc.
Walter von Kleinsmid

"Mother Goose" on step in this photo sent from Walter von Kleinsmid to Mr. Igor Sikorsky in 1958. Kleinsmid was co-owner with Dick Probert of Avalon Air Transport. (Proberts)

nance for Panagra and the chief of maintenance for Braniff. Both of these people had years of experience maintaining Pratt & Whitney model R-1830-92 engines when these same engines were the engines that powered the DC-3s when almost all airlines were using DC-3s. When I advised these two mechanics exactly how the two engines which failed in flight acted prior to their quitting altogether, they both agreed that the problem was spark plugs. Now I obtained my CAA Airplane & Engine license in 1943, and I, at that time, had very little experience on P&W 1830-92 engines, but, it didn't seem reasonable to me that all 56 spark plugs in these two engines would stop working in the two minutes time between when I first noticed roughness in the engines and when they quit altogether. I considered the difference between their experience and mine and went along with their theory.

"They installed 56 new spark plugs in the two engines which were giving us trouble. We thereafter flew the airplane and, surprising enough, all engines ran quite normally. In less than ten minutes flight time I decided that the problem was behind us and landed and ordered gasoline. About this time, I was thinking that I didn't know as much about aircraft engines as I thought I did!

"We taxied the aircraft as close to

shore as we dared, threw out the anchor and hoped it would hold. It did. The fuel company on their first attempt, sent out a regular gas truck with enough hose to reach across the beach and across the ocean bottom to the airplane. With the aid of a local gentleman named Eduardo, whom we had hired to furnish a boat and the muscle to row it, we got the gas nozzle to the airplane and into the gas tank opening. After pumping about 200 gallons of gasoline into the airplane tank, the pump on the fuel truck failed. As there was nothing else we could accomplish, we scrubbed the fueling operation for the day. The next day, May 12, the fuel company sent out a larger tank truck along with a pumper. The pumper was required as a result of the great length of gasoline hose that carried the fuel from the fuel supply to the airplane. The operation was successful and we filled all tanks to capacity; 4200 gallons.

"The operations manual for this airplane, which, at that time I did not have, advises, that for structural reasons, to limit the fuel load to 3800 gallons. So we were 400 gallons or 2400 pounds too heavy on the gas load. In addition, we had loaded all the spare parts and equipment that the previous owner had left at Ancon. Obviously, because of its size, we could not bring the beaching gear with

us, these had to be shipped to Long Beach via boat. I had planned to make the ferry flight from Lima to Long Beach non-stop. Considering the fuel capacity of this aircraft such a flight was entirely feasible. The most direct route would follow the South American west shoreline in a north-westerly direction starting at Lima and proceeding to a point just north of Chiclayo and from there over 1,200 miles of open water to the west coast of Mexico, possibly to Acapulco. From Acapulco it is just a case of following the shoreline to Long Beach. Because the aircraft had no navigational equipment which was operational, I decided to navigate the over-water portion of the flight using celestial navigation.

"During World War II I was a flight captain with a civilian portion of the Air Transport Command. The contractor to the military was Consolidated-Vultee Aircraft Company. CVAC set up a separate division which they named Consairway. Consairway required all captains flying their transport command routes to be able to replace any crew member that became incapacitated. I therefore had to learn celestial navigation. Fortunately I have been blessed with considerable ability in mathematics and mathematics is what celestial navigation is all about. Under the watchful eye of my navigator, I would navigate

"Mother Goose" at the ramp at Long Beach, Ca. (Proberts)

every other leg of most flights. Before I left for Lima to ferry the Sikorsky to Long Beach, I went to a friend who was operating an aviation ground school where I took a brush-up course on the latest methods used in celestial navigation. Such navigation had been considerably simplified since World War Two. Experience told me that night-time use of celestial navigation was much simpler and much more accurate than trying to "run down a sun line" in the day time. I therefore planned my time of takeoff just a little before sundown, which would allow me to follow the shoreline north to a place where I would lose the shoreline and be entirely over water. I planned to be at this place when it became real dark, giving me a large choice of stars to navigate by.

"The takeoff was no problem, we climbed normally to 8,000 feet and reduced power to cruising power. By this time, it was dark. Very shortly after reducing power my co-pilot, Sully Sullens, pointed to the tachometers for number two and number four engines. These two tachometers indicated extreme roughness in the engine operations. This was followed by backfiring that lit up the sky. Within three minutes, these engines were delivering absolutely no power. I had no choice. I had to return to Ancon. This was a bit of a harrowing experience; an overloaded airplane with two of its four engines not operating, a strange place, a flight crew that had very little experience working together, a strange airplane in which I had less than one hour experience, a harbor in which many fishing boats sat at anchor without lights of any kind, night time with a thin overcast with low scud below, no way to obtain wind direction, and no way to determine height above the water.

"When I arrived back at Ancon I was low, possibly less than 1,000 feet. I circled the small community hoping to attract someone's attention. I did—the police. After circling the community, I flew north planning a 180 degree turn for a landing to the south directly into Ancon harbor. Using full flaps, I set the power and trim controls for a 200 foot per minute rate of descent with the aircraft in level position. During the ten years that I flew this aircraft thereafter I never made a better landing. I had a gold horseshoe you know where.

"So, I was on the water and luckily didn't run into a fishing boat; now all I had to do was to find our mooring and get the airplane attached to the mooring.

As stated before, taxiing with an inboard on one side of the airplane and an outboard on the other side is somewhat difficult if you are water-borne. Where is our mooring? In the distance we observed some lights waving up and down and around in circles so we assumed someone was on our mooring signaling to us. When we finally arrived at the mooring we found several police officers who helped us tie up the airplane. When we got ashore, they escorted us to the police station where they arrested us for entering Peru with no permit. Fortunately, Ruzakowsky had married a Peruvian girl, had been living in Lima, and learned to speak the language quite fluently. After much yelling and arm waving between Ruzakowsky and the police, we were set free. Also, fortunately, our driver, which we had hired to drive us around Lima as required, was still in Ancon when we returned. He drove us back to the country club. John Forester was still at the bar drinking to the departure of his airplane.

"At that time, we were still working on plan B. We hadn't bought the airplane yet. When John saw us he became sober real fast. He was most insistent that we stay through the next day to see if his mechanics could remedy the engine problems. The next day, after considerable discussion with the two mechanics, they decided that the problem must be "sticking valves." They suggested, and I agreed, to fill the carburetor alcohol tanks with mystery oil. By doing so we should be able to inject this highly penetrating oil directly into the cylinders while in flight and if any valves were sticking this method should loosen them up. We flew the airplane again with the same results. When the engines started to act up, we shot mystery oil into the cylinders through the carburetors This didn't help a bit. By this time, I'd fallen in love with the airplane and began to see where I could add additional seats. The galley compartment

located directly under the flight deck had been set up as an office with filing cabinets and other office equipment. I could put eight seats in that compartment. At this point, it appeared to me that getting the two engines to run as they should was beyond the capabilities of John's two mechanics. As I saw it, the worst possibility that we faced in getting the airplane to Long Beach was the total replacement of the two bad engines. The engines ran real good when they were cold so it had to be something simple to correct.

"Being a business man, I took advantage of a bad situation for the sellers. Here they had a four-engine seaplane for which I was the only likely purchaser in the world. It was in the water with no pilot except me to get it ashore. If I left and went back to Long Beach, what would they do? On May 16th, I offered Forester $65,000 for the airplane and if that was unsatisfactory, I informed him that I would be on my way home. John phoned his principals in Baltimore, and, after much consternation, they accepted my offer. I phoned Walt from the country club where these negotiations were taking place, and advised him of the new price of the airplane and what bank to send the money to.

"Within a few hours Walt called me back to advise that he had been with the FAA and that the FAA would require a hangar for the airplane in which to accomplish maintenance and, of course, we would have to have a ramp to get it out of the water. This had a real dampening effect on my spirits for I knew that there was no way that we could afford to build a hangar for an airplane with a 124 foot wingspan. Where would we build it? Where could we purchase beach property on which to build a ramp? I, therefore, had to call the whole deal off. The next day, while we were all still together, Walt called to advise that the FAA had waived the hangar requirement if we operated the airplane only during the

Passengers board *Excambian* for the trip from Long Beach to Catalina. (Proberts)

summer months. Also, Walt advised that he had been with the U.S. Navy and they agreed that because we were a contractor to the Navy in the development of the Polaris missile, we could use the Navy's ramp at the old Reeve's Field in Wilmington and that we could also store the airplane on the Navy property during the winter months. I immediately met with Forester to advise him of the latest development; however, I played up the necessity of buying shoreline property on which to build a ramp and thereafter came up with an offer of $55,000 for the airplane. Forester called Baltimore and again they accepted my offer.

"When I called Walt with the latest price information, he said "If I leave you down there another few days maybe they will give it to you." At this time it appeared that we had a definite deal and I therefore requested Walt to send down our chief mechanic with enough tools to overhual a P&W 1830-92 engine. I had

expected that to get Bert Taylor, our chief mechanic, a passport, the necessary visas and other paperwork, and get him to Lima would take at least one week. Because John Forester's mechanics stated that the trouble we were having with the engines may be "sticking valves" and because I didn't expect to see Bert for at least a week, I decided to remove the cylinders from one engine and have the valves checked. This was possible only because Faucett Airlines, Peru's largest airline, had their maintenance facilities in Lima. Faucett Airlines was not in the habit of taking in maintenance work from outsiders. Their maintenance facilities were for their own use only.

"In connection with making the sale of the Sikorsky to Avalon Air Transport, the liquidators who were handling the sale, through their employee John Forester, had made contact with one Elwood Alexander, an ex-Pan American Airlines publicity agent, who ran a travel

agency in Lima called Dasatour. Elwood Alexander was with John Forester when I first visited Ancon to look at the airplane. In some small way, he participated in the sale. Elwood had been in Lima for quite a few years and was acquainted with numerous people in high places. He knew, personally, the maintenance manager for Faucett Airlines and it was through his connection that Faucett agreed to help us. Elwood Alexander, when the Sikorsky deal was behind us, would play a very big role in AAT acquiring two Grumman Gooses from the Peruvian government.

"While Walt was getting Bert Taylor's passport, visa, etc., and getting him on his way to Lima, I, with the help of Ruzakowsky and Sullens, my co-pilot, commenced removing cylinders from the number two engine. On May 18th, we removed three cylinders. On the next day we removed three more. As we got around toward the bottom of the engine it became extremely difficult to reach the cylinder hold down nuts from any posi-

A wonderful photo of *Excambian* at the dock in Avalon Bay. The famous Avalon Ballroom is in the background. (Proberts)

tion on the wing work stands. We therefore constructed a floating work stand. We made this floating work stand out of four old life rafts which were of the type carried aboard most vessels during World War II which were subject to being sunk by German submarines. Fortunately, there was a good supply of these rafts available to us. For a floor on this floating work stand we purloined sections of an unused dock and parts of an old pier. This floating dock worked real well except that the water would not co-operate. As the airplane went up, the work stand went down, thus giving our arms a real good workout while holding a wrench overhead trying to get the wrench on a cylinder hold down nut. On the day that we built the floating work stand we delivered six more cylinders that we had previously removed from the engine to Faucett maintenance.

"The following day we contacted Faucett maintenance who advised us that the cylinders were in poor condition and that the clearance between the valve stems and the valve guides were excessive and therefore there was no possibility of the valves sticking. On the same day, May 21st, we removed five more cylinders and the next day we took them

to Faucett. While at Faucett, we picked up six more cylinders on which they had overhauled the valve mechanisms. Bert Taylor arrived in Lima on the same day. We met Bert at the airport. He got off the plane with a request to see the Sikorsky. Every time that I had made the trip to Lima via DC-6, I was so tired when I got there that all I wanted to do was sleep. Not so with Bert; he said he slept all the way down.

"While driving from the airport in Lima to Ancon, Bert requested that I advise him in detail just how the engines operated. This I did to which he replied "That's mags Dick. That is all it can be." We kicked that back and forth for a while when I came up with a statement to the effect that this airplane has four engines and that each engine has two magnetos. What are the chances of four magnetos quitting at exactly the same time and that the four magnetos that quit are on just two engines? His reply was "I'll stake my A&E on it."

"The next day, after Taylor arrived we started making progress on the engine problems. With three licensed airplane and engine mechanics, Taylor, Ruzakowsky, and myself, with Sully, who was also no slouch at engine work, we

started showing signs of getting the airplane to Long Beach. On this day, we installed six of the 11 cylinders that had been removed. We also, on this same day, picked up the balance of the cylinders, five, from Faucett who had reworked the valves. These cylinders were installed the following day.

"Normally it shouldn't take so much time to install cylinders but these engines had been equipped with fire extinguishing equipment which involved stainless steel pipes. These pipes had absolutely no give to them and, as a result, the attachment of the pipes to the cylinders had to be made as the cylinders were installed and fitted to the crank case. About six hands were needed as each cylinder was installed. It took the next four days to finish up the work necessary to get number two engine assembled and running. It must be remembered here that we were working in much less than ideal conditions. We were working outdoors, the wind blew continuously, causing a good chop on the water with light swells, we were working from a floating work stand which, as noted before, was extremely difficult to work from, if we dropped a wrench or a part it went to the bottom of

Dick Probert exiting *Excambian* through the rear passenger hatch. (Proberts)

the ocean with no chance of retrieving it, when we had to go ashore for anything, it involved calling for Eduardo and his boat to row us ashore. This was a long ways from working on solid ground inside a hangar.

"The next four days, May 25th through May 28th, we spent finishing up the work necessary to get number two engine assembled and running; removing magnetos and carburetors, taking them to Faucett for checking, reinstalling after checking, etc. Although Taylor's diagnosis of our engine problems was probably correct I wasn't thoroughly convinced. As a result, we removed both magnetos from number two engine and took them in to Faucett. Faucett had one large, whole room in their maintenance facility devoted entirely to magnetos. Through our interpreter, I requested that the magnetos be installed on their magneto test stand and I requested that they run the magnetos for two hours. At the end of one hour, the person in charge of the magneto

shop came over to the test stand and shut it down. Again, through our interpreter, I told the gentleman to let it run for another hour. From that point, the conversation went like this, "I can't," "Why not?" "Its lunch time." "Leave it run, we'll watch it." "Can't" "Why not?" "We shut the power off for the entire building to be sure nobody works during the lunch time." Four hours later, after "siesta" we managed to get the magneto to run for two consecutive hours.

"At the end of the two hour test run, the magneto test stand indicated that the magneto was functioning perfectly. I asked Bert, who was standing next to me, what he thought about the test. Bert told me to put my hand on the magneto and tell him whether it was hot or cold. I did and found it cold. Bert explained that the magneto, when attached to the engine case, absorbs the engine heat through the magneto attachment pad and, apparently, in our particular case, it takes 15 to 17 minutes of engine operation to get the magnetos hot enough to

break down the magneto coils. He said that if we could put new coils in the magnetos they would probably operate satisfactorily. We requested Faucett's magneto shop to overhaul all four of the magnetos that we felt were giving us trouble. At that point the conversation went: "We can't overhaul those magnetos." "Why can't you?' "We have no parts for Bosch magnetos" "Bosch magnetos are a widely used magneto, why don't you have any parts for them?" "Because Bosch magnetos don't work in Lima!" This was the first time that I heard this statement. What possibly can Lima have to do with the operation of Bosch or any other brand of magneto? The Faucett magneto shop was filled with Scintilla magnetos and Scintilla magneto parts, but no Bosch magnetos or magneto parts. There were two Scintilla magnetos in the spare parts box that came with the airplane. Neither had the correct "cams" for the engines that powered the VS-44A. Faucett could change these cams and install the correct ones in our two spare magnetos. We asked Faucett to do this.

"In the mean time we installed all of the magnetos back on the engines and again flew the airplane. This was on May 29th. Engines number two and four still became rough after 15 minutes of flight. It, at that time, became clear that we must have new magnetos to replace all of the Bosch magnetos that were installed on engines number two and four. We had the two Scintilla magnetos on which Faucett had changed the cams. That left us short two magnetos. The driver we had hired to take us where we wanted to go and who also acted as our interpreter suggested that we try the Peruvian Air Force parts department. When we arrived at the armory, our interpreter advised the sergeant, who seemed to be in charge of the parts department, what we were looking for. Bert Taylor felt that if we could get two new coils we could eliminate the problem. Our interpreter asked for new coils and, after a trip to the rear of the large parts room, he threw two coils on the counter. Taylor advised that they were the correct coils, although they were extremely dusty, indicating that they had been in storage for a considerable length of time.

"At the time, the sergeant threw the coils on the counter, he asked in perfect English, "Is this what you are looking for?" We advised him that the coils were exactly what we were looking for. To which he asked if we would be interested

Damaged float on the flight back to the US from Peru. (Proberts)

in obtaining two new Bosch magnetos. I asked to see the new magnetos after which he produced two Bosch magnetos of the same style and design as were used on the VS-44A, only they were not new. They had a crackle finish which no new magnetos possess. I turned the magneto, which I had in my hand, upside down and on the bottom was a decal which read. "Overhauled by the Babb Company." I questioned the sergeant about this to which he advised "You are right, those magnetos were removed from an overhauled engine which we purchased from the Babb Company." I asked him why they were removed from a freshly overhauled engine. His reply, "Bosch magnetos don't work in Lima, we installed Scintilla magnetos in their place."

"The sergeant asked $190 each for the two Bosch magnetos until I showed him that they were not new. We bought them for $90 each as used parts.

"We now had four magnetos for the two engines that were giving us trouble. The date was May 31. We went to Ancon and installed three of the magnetos. On this date we took the radios to Faucett for a check. The next day, we installed the last of the magnetos and flew the airplane a total of two hours and 15 minutes. All engines ran perfectly. The following day, June 2nd, we again gassed the airplane, reinstalled the radios and departed for Long Beach—we thought!

"It seems that the previous pilot who flew the airplane to Peru, Hugh Wells, was hesitant about landing on the Amazon at Equitos. Too many floating logs. He, therefore, prepared the airplane for the possibility of striking something on the water while landing and tearing a large hole in the bottom. This airplane was designed to stay afloat with any two compartments flooded. Being curious, I had opened the bulkhead doors, which formed parts of the floors in the passenger compartments, to look inside the hull. In there were sheets of aluminum, shears, rivet guns, underwater lamps, joint sealing compound, drills, scuba equipment, etc., to repair a hole in the bottom hull skin. What I didn't see was a 300 pound roll of canvas that was stored in the rear baggage compartment.

"During flight tests, with 4200 gallons of gas aboard, Ruzakowsky, our flight engineer, had noticed that the flight level indicator, which he had near his engineer's panel, was indicating a slightly nose up attitude when we were flying level. Without considering the weight of the fuel we were carrying, which made it fly slightly nose up, he decided that there was too much weight in the tail and, without my knowledge, moved the 300 pound canvas from the tail compartment to the nose compartment. This of course made the airplane completely out of balance nose heavy.

"For the two weeks or more that we had been working on the engines there had been a brisk breeze with a subsequent moderate chop on the water. On the day that we left there was no wind blowing and the water surface was glassy. I made five tries to get the airplane on the step but each time I tried the nose would come up but not enough to get on step. each try ended up with the nose falling back into normal taxi attitude and with all four engines at full power. The propellers took a real beating with water contact. After the five tries at takeoff without success, I decided to abandon a takeoff. I therefore started taxiing towards our mooring expecting to put the airplane on the mooring for the night. On the way to the mooring, and while quite close to shore, I noticed a tiny ripple on the water which to me, indicated that a very light breeze was blowing. It was blowing parallel to the shore line. I lined the airplane up into the wind and decided to try once more. I opened the throttles, pulled on the wheel and she went up on step. I figured that I had made it. The faster we went on the water the more the nose wanted to dig in and I had to call for help by the co-pilot to hold the nose up. We were both pulling hard when she lifted from the water because the shoreline, which was on my left, made a 90 degree turn to the right just ahead of me. There was a hill on the shoreline, which I wasn't about to get over. I had to make a turn to the right. At that time I had been flying Grumman Gooses for about four years. You can take a Goose off the water and immediately make a very steep climbing turn. The Goose had fantastic performance. In making my right turn in the Sikorsky I had not considered the 124 foot wingspan and I put the wing down far enough to catch the right wing float in the water. I felt a very slight jar in the airplane but everything was going well. At about 300 feet of altitude, Bert, who was sitting just behind me, tapped me on the shoulder and pointed to the right wing float, or what there was left of it.

"The wing tip float on a Sikorsky VS-44A measures 13 feet long and three feet in diameter. The float, which contacted the water, broke exactly in half upon water contact. The two halves lined up alongside each other and, with the float wires holding them, went up into the trailing edge of the wing, through the wing so that one half of each of the half floats protruded above the wing and the other half hung below the wing. This "junk" struck the wing trailing edge at the junction of the wing flaps and aileron which left me with no aileron control or flaps. It did leave me with an obstruction which measured 6 feet by 6½ feet located near the wing tip which the airplane had to push directly into the wind. With this obstruction in the wing the aircraft cruised at 114 mph instead of the normal cruise of 185 mph. Fortunately, the autopilot worked very well, holding one rudder pedal ahead of the other rudder pedal about five inches

Excambian at Avalon. (Proberts)

to compensate for the drag imposed by the wrecked float.

"It was decided among the crew that to return to Ancon would be taking a chance on losing the airplane. Whereas we had sufficient fuel to make Acapulco, that's where we would go. We flew all night and landed the next day, at 9:30 a.m., in Acapulco. During the flight to Acapulco, we burned all the fuel from the right wing tank as well as from the main tank leaving all the remaining fuel in the left wing tank. As a result of this fuel management when we landed, being left wing heavy, the left float went in to the water, and, so far, we were not in too serious trouble. Landing a seaplane, with one float visible, in Acapulco Harbor, and taxiing into the yacht marina is not an everyday occurrence in Acapulco and, as a result, we had no problem getting ashore. We had lots of help. When ashore, we went to the police station to explain our entry in to Mexico without prior permission.

"We had no trouble there, except, the Mexican equivalent of our FAA wanted to see the airplane. He spoke not a word of English. We hired a boat and took him to see the airplane and when he saw it he exclaimed "impossible." This word means the same in Spanish as it does in English. While in the police station explaining our problem one of the policemen advised us that we should contact a John Wemburg, who they said, could help us. The name Wemburg didn't sound very Mexican so I figured he must be some other nationality. Before this day was over we had made contact with the operator of a water ski school who had considerable dock space on the city side of Acapulco Bay. We made arrangements with him to dock our airplane at his docks and, by using two large anchors, which we had on board, anchored the airplane in place with the damaged wing over the land where we could work on it. After making sure all was well we walked into town and found a fair hotel where we holed up for the night.

"The next morning, I met John Wemburg in a small restaurant across from our hotel. John was a tall, blonde, blue-eyed man whom I took to be a German national. He spoke perfect English.

"He was a pilot and had his own airplane with which he did a certain business, about which I did not question him. I told him about the VS-44A and that I needed a small amount of fabric, dope, thinner, and rib stitching cord and that I intended to pay for his services. I told him that I also needed about 1,800 gallons of 100 octane aviation fuel and the equipment to pump the fuel into the airplane's tanks. He advised that the fuel might be a problem.

"Later, he took me to several hardware stores where we bought about 200 feet of new garden hose and an old hand-operated pump. Next we headed for the wholesale fuel distributors, where I was advised that it was against the law to haul gasoline through the city of Acapulco and that I probably would have to bribe somebody to do it. Finding someone willing to do this took quite a bit of running around. John obtained the rest of the supplies that we needed and we took everything to the airplane where the balance of my crew was removing the damaged float.

"About this time, 150 small Mexican boys along with the Mexican FAA person had gathered to watch the proceedings. I motioned to the Mexican FAA person to come with me and he did. When I was out of earshot from anybody, I said to him in my best high school Spanish "I will pay you 200 pesos if you O.K. my seaplane to takeoff with the authorities in Mexico City." To which he put his arms around me and said "Me Amigo-My friend." That man would not let me out of his sight. He used the government car to take us to and from the airplane. He had dinner with us. He had breakfast with us. He was with us at all times except in bed. When I told John what I had done to get rid of the FAA problem and how much I had promised to pay him, John said "God, you've given him a year's pay!"

An excellent view of the tail and horizontal stabilizer of *Excambian* while in service with Avalon Air Transport. (Proberts)

"The next day, which by that time was June 5th, the gasoline was delivered and we were still on the way to having the necessary repairs made to the flaps and aileron. The float damaged the flap and aileron in such a way that the aileron touched the end of the flap just enough to keep the aileron control from working freely. With all the small boys that were continually with us I figured out a way to get the 1,800 gallons of fuel in to the tanks without the crew doing the pumping. I offered a prize of $5 to the boy who, unassisted, could pump a 50 gallon drum dry in the least amount of time. We had a lot of takers to that one. One boy did it without stopping. When one arm wouldn't take it anymore he pumped with his other arm. Needless to say, the gasoline went in to the tanks with great haste and with no trouble to the crew.

"With the flap and the aileron repaired and all the gas we needed in the airplane, we were ready to go. All we had left to do was to get a nod from the FAA inspector. He got in the airplane, in the cockpit, and moved the aileron control its full limits. He noticed the hang up when the aileron lined up with the flap but said nothing. He got up out of the pilot's seat, said O.K. and I handed him 200 pesos. At that time the exchange rate was 7 pesos for one dollar. The next day we left Acapulco.

"On Thursday, June 6th, we prepared to leave for Long Beach, Cal. After the anchors were aboard, I fired up the engines and taxied to some open water where I let the engine oil warm up while the aircraft made a large circle to the right. The circle to the right was necessary in order to keep the left float in the water. The right wing float was still on the beach at Acapulco in two pieces. We had cleaned off all the float attachment gear, the struts and the wires. While the aircraft was still turning to the right, warming up the oil, I pulled the power to idle on the two right engines as well as the left inboard, and thereafter proceeded to check out the left outboard engine. The magnetos, temperatures and pressures, are checked at slightly above cruising power, and therefore, when checking an outboard engine, the airplane turns sharply away from the engine being checked. I checked all the engines except the right outboard and when I did check that engine I forgot about the right wing float being off and when the aircraft began turning sharply

to the left the right wing went down and touched the water. That didn't bother me for I assumed that a sharp turn to the right would pick the wing up and put the left float in the water and we could go on from there.

"Unfortunately, with the two left engines at full power the right wing would still not come out of the water, in fact, it went further under the water as time went on. There was nothing left to do but put as much weight as possible on the left wing at the tip. All four of us left the cockpit and shinnied up the left wing all the way to the tip. I was sure that with all of us that far out on the left wing, the right wing that was partially under water would come out of the water. It didn't. Next, a line was attached between the wing's tie down ring and a motor boat. This was accomplished by holding Sully by his ankles and lowering him head first over the wing leading edge at the place where the tie down ring was attached to the lower surface of the wing. By this time, the right wing was submerged to the point where the outboard engine propeller was well down in the water. With the engine cowling about to get wet, our crew, sitting on the wing tip, were quite a distance over the water. Luckily, this did the trick and the motor boat was able to move the plane back toward its center of gravity. However, when the crew hurriedly left the flight deck to go to the wing tip the engineer had omitted turning off all gas valves and, as a result the gasoline had siphoned from the left wing tank through the fuel system across the

airplane in to the right wing tank. In addition to this, the right wing was holding considerable water.

"The crew sat on the left wing tip for almost one hour causing the gasoline in the right wing tank to flow into the left wing tank and also allow the water to drain out of the right wing. At the end of an hour's time we were able to enter the flight deck without the right wing dropping down into the water. I fired up the engines, taxied to open water and took off. It flew beautifully, even without one float. As a result of transferring fuel across the airplane we had absolutely no idea what tanks held how much fuel. We knew that, originally, before the wing tip went in the water, we had sufficient fuel to make the trip to Long Beach, but what we didn't know was how much fuel went overboard when the right wing was in the water. After reaching cruising altitude, we shut off all fuel except the right wing tank. That tank ran for about two and a half hours before it ran dry. This left us with the gas that was in the left tank plus the gas that was in the reserve tank. The reserve tank, when everything is normal, holds 125 gallons which should run the engines for about 40 minutes. We had no idea just how much spilled out of that tank when the wings were so badly out of level.

"Long before we reached Ensenada, Mexico, I was continuously looking for bays and backwaters to land in should this beautiful airplane run out of fuel. When we approached Ensenada I was tempted to land there for they have a beautiful bay. I knew that in a few

minutes I would be in San Diego and in the United States. At San Diego, I switched on the reserve tank for I wanted all the fuel in the airplane in the left wing tank, so that, after landing, it would come to rest with the left float on the water, I hoped. At San Diego I decided to keep going for there are lots of places where I could land between San Diego and Long Beach. I made Long Beach with 15 minutes of fuel left in the left wing tank.

"And so, the VS-44A got to Long Beach where it was immediately moved to the old Reeve's Field where the Navy was, among other things, working on the Polaris missile. Without my knowledge, Walt had placed complete insurance coverage on the airplane before I had left Ancon. The insurance company gave the necessary repair job to Frank Nixon, who ran a complete repair shop at the Van Nuys Airport. Frank made the necessary repairs in two or three weeks, after which we put the airplane in service to Catalina and San Clemente Islands. And this began another odyssey for the VS-44A which lasted for ten years."

The configuration of the aircraft, as it was acquired by AAT was essentially as it was when it left the factory some 15 years earlier. The passenger compartment was set up to seat 32 passengers on day time flights and 16 passengers on overnight flights which required berths. One of the first modifications made by AAT was to remove the berths and reconfigure the passenger compartment to seat a total of 47 passengers. Of course, this required a supplemental type certificate approved by the FAA. The other major change undertaken by AAT approximately 18 months after acquiring the

aircraft was to remove the flight engineers panel. AAT, like most airlines at that time, had to be cost conscious to survive. The expense of having two pilots, a flight engineer, a radio operator, and a navigator as required crew on the 27 mile trip back and forth to Catalina was financially obtrusive. By relocating essential controls from the flight engineer's panel to within reach of the pilots, AAT was able to convince the FAA to issue a supplemental Type certificate to cover this modification. The success of this modification ultimately enabled AAT to eliminate the flight engineer position. The positions of navigator and radio operator were also subsequently eliminated, enabling AAT to operate the aircraft with a flight deck crew of two pilots only.

Additional Type Certificates were issued to AAT to cover the addition of a FAA-required anti-collision light on the vertical stabilizer, and a baffle for the main gasoline tank which would serve to catch any fuel which leaked from the tank and purge it overboard. As saltwater erosion took its toll on the original equipment "Baby Hydromatic" propellers and it became increasingly difficult to locate replacements, alternate replacements became a necessity. A model of propeller that was standard equipment on the PBY were chosen as replacements. These propellers were heavier and their use caused a slight degradation in performance.

In 1959, #41881 became an actress. Dick Probert recalls her debut in a Hollywood movie; "In April of 1959 a motion picture entitled *The Gallant Hours* was made. The motion picture company, MGM, advised that the entire

aircraft had to be painted a dark navy blue for the picture. This movie was to be the World War Two story of Admiral William "Bull" Halsey and was to be shot at San Diego. After the picture was completed, the studio, who had contracted to remove their paint, was unable to accomplish that task. This made it necessary to repaint the entire aircraft. The repainting of the airplane caused the fabric on the wings to shrink to the point where it tore loose at the corners where it was attached. Our mechanics repaired the fabric enough to see us through the coming summer and after summer was over we removed all the wing fabric and covered the entire wings with metal. This was a big job. There was only one plater in the Los Angeles area who had vats large enough to anodize the large sheets required for this installation. When this work was complete a flight test was required by the FAA. I took the airplane up with the FAA on board and dove it to its maximum speed allowable to prove that the metal would stay on and in doing so completed the certification requirements for the metal installation."

In 1960, Avalon Air Transport became a U.S. Certificated Scheduled Airline and as such operated under a certificate of Public Convenience & Necessity issued by the Civil Aeronautics board. This change required AAT to establish emergency exits. This was done by modifying the large passenger windows and labeling them as "emergency exits." The FAA required AAT to demonstrate the suitability of this modification as well as the emergency lighting associated with it by conducting an emergency egress drill. The drill was successfully completed and the FAA approved the modification.

Avalon Air Transport owned #41881 *Excambian* for approximately ten years. In that time, the aircraft made the 27 mile, 12 minute crossing between Catalina and Long Beach 8,172 times! An estimated 211,246 passengers were carried during that ten-year period. An additional 68 round trips were made between San Clemente and Long Beach under contract to the U.S. Navy. Additional charters were made to San Francisco and points in Mexico. Consequently, Dick Probert had more VS-44A take-offs and landings than all other Captains combined. AAT's chief stewardess, Nancy Ince Probert, made 7,178 crossings to Catalina and back in *Excambian* as well as being on all other charter flights.

Antilles Air Boats

Excambian Sold To Antilles Air Boats

On February 14, 1967, Dick Probert celebrated his 60th birthday. The celebration was, unfortunately, tarnished because the FAA had decreed that pilots 60 and over could no longer fly for scheduled airlines. Dick knew that if he remained the owner of #881, sooner or later, he would succumb to temptation and get into trouble with the FAA. *Excambian* was sold for $100,000 to Charles Blair. Blair owned Antilles Air Boats which operated under an Air Taxi Certificate.

Above: Charles Blair and his wife actress Maureen O'Hara Blair on the flight deck of a Pan American Airways Boeing 707. After Captain Blair's accidental death in 1977, Maureen assumed management of Antilles Air Boats which had a fleet of 22 planes, 26 pilots, and a large measure of issues which needed to be dealt with. As the first female CEO of a major airline, she effectively managed without any prior experience. (Bob Quinn)

Below: *Excambian* in San Juan, Puerto Rico during her last year in service flying for Charles Blair and Antilles Air Boats. A Sikorsky HH-52 in service with the U.S. Coast Guard is to the right. (Marcus Blair)

Charles Blair entering the cockpit of his P-51 *Excalibur III*. Blair made the first single-engine, solo flight over the Arctic in this aircraft. The National Air and Space Museum restored *Excalibur III* and it is currently on display in a museum in California. (Christopher Blair)

meet passenger demand between its St. Thomas and St. Croix hubs now required an aircraft with the capacity of the VS-44A.[16]

So in January 1968, Charlie Blair with the assistance of Dick Probert, transported the *Excambian* from Long Beach to the U.S. Virgin Islands. Fuel stops on the ferry flight were Eagle Mountain Lake near Ft. Worth, Texas, and Miami, Florida. Blair's 17-year-old son Christopher was in charge of arranging the fuel stops and joined the flight in Ft. Worth. For the remainder of the flight, he was to also serve as "unofficial" auto-pilot. As Charlie was quite busy with the day to day operations of Antilles Air Boats, he requested that Dick stay on long enough to check out Antilles' chief pilot on the *Excambian*. For the next year, the aircraft performed her role quite effectively. On January 10, 1969, however, the aircraft was involved in an unusual incident that was to end her long service life.

There has been considerable controversy regarding the final flight of the *Excambian*. Recently, contact was made with Mr. Ted Pfeiffer, who was purser onboard the aircraft on that fateful day. Mr. Pfeiffer still resides in the Virgin Islands and his memories of that day and his experiences with the VS-44A are as

This technicality allowed Dick to assist Charlie in ferrying the VS-44 to the Virgin Islands and to continue on there for some time to check out the other Antilles pilots in the aircraft.

It was fitting that *Excambian* was sold to Antilles Air Boats. Charles Blair had been the first pilot to fly the VS-44A as test pilot on the Excalibur. At that time, he was the chief pilot for American Export Airlines and performed double duty, with AEA's blessing, as Sikorsky's chief test pilot on the VS-44A. As he had expressed to Fran Wallace in 1942, he thought the VS-44A was an excellent aircraft. He later contracted with both the Aviation Exchange Corporation and Skyways to check out their "44" pilots. He had established many world records with the "Flying Aces." It was fitting, therefore, that as owner of Antilles Air Boats, "Charlie" would provide the last operational home for the last of the "Flying Aces."

Charles Blair started Antilles Air Boats in 1964 with one Grumman Goose to provide a scheduled air taxi service

between St. Croix and St. Thomas. Over the next several years the convenience of downtown to downtown service really caught on and he was soon providing service also to Puerto Rico, St. John, and Tortola. By 1967, the AAB fleet had grown to five Grumman Gooses and to

Blair receiving the Collier Award for the record-setting Arctic solo flight from President Harry S. Truman. (Christopher Blair)

Actress Maureen O'Hara Blair continues to maintain a keen interest in aviation. She has made numerous visits to the VS-44A restoration hangar to check on the progress of the work. Her lively spirit and devotion to her late husband and "his" aircraft have made her a favorite of all the volunteers.

clear as can be. Ted recalls that day, "We were returning to St. Croix and as usual were making our approach from the west. Captain Bill Sorrens was the Pilot in Command and he essentially had two co-pilots, a young, somewhat inexperienced, but nevertheless talented co-pilot and a very experienced flight engineer who had recently retired as a captain with a major airline. As we came off step, the number four engine failed. Both co-pilots noticed it immediately. Captain Sorren was busy controlling the aircraft and evidently did not notice the problem. As we entered Rollover Gut, Bill revved the two outboard engines to compensate for drift caused by the wind or tides and, of course, with number four engine out, the increased rpms on number one engine caused the aircraft to lurch to starboard and she went aground. Bill immediately feathered number one engine, and the aircraft promptly sank. No one has ever figured out why neither co-pilot warned Bill regarding the engine outage. Fortunately, the water was only four feet deep and as the aircraft quickly settled to the bottom, only the passengers in the very first compartment, which was actually the galley, got their feet wet. That compartment is two steps down from the "real" passenger compartments. The aircraft was quite close to shore and as a matter of fact, the starboard wing was close enough to land to allow a very attractive young lady named Molly to crawl out on the wing with food and beverages as we pondered our situation. None of the passengers or crew was injured and with Molly's help we were actually pretty comfortable. We stuffed the gash in the hull full of rags and other cloth and eventually managed to pump out enough water to taxi the aircraft to the ramp. She was too heavy to taxi up the ramp on her own power so we winched her up. I later built a concrete pad for her

and there she sat. She became a local landmark." Mr. Pfeiffer also recalled that he loved flying in the VS-44 so much that he could not wait to get to work in the morning and, in fact, would have worked for free!

Shortly after the mishap at St. Thomas, Charles Blair consulted with Dick Probert regarding the possibility of Avalon Air Transport's maintenance crew refurbishing and repairing *Excambian*.

Because of the extensive corrosion on the airplane, the price to complete repairs to an airworthy condition was prohibitive and *Excambian* was relegated to becoming an item of curiosity parked for the next six years near the launching ramp in St. Thomas.[17] Sever? the aircraft were contemplate ing one where *Excambian* w been the centerpiece for a the rant, but none was ever imple

Excambian Restoration

The six years of sitting idle at the ramp took its toll on the aircraft. The hot, humid days and the salt spray accelerated the oxidation of the airframe and aluminum skin. Charlie Blair and his wife, the actress Maureen O'Hara, realizing that it would be economically unfeasible to refurbish the aircraft to operational status, and concerned that time was taking its toll on the aircraft, decided that it would be best to donate *Excambian* to a museum.

Doing so would allow the aircraft to be restored for static display and preserve for future generations the romance of the flying boat era. In 1976, the aircraft was offered to the New England Air Museum, but unfortunately the museum could not arrange to have the aircraft transported to Connecticut. The Blair's subsequently donated the aircraft to the Naval Aviation Museum at Pensacola, Fla.[18] Unfortunately, because the *Excambian* was not officially a former U.S. Naval aircraft (despite its service for the Naval Air Transport Service during WW2), it did not receive a high priority for restoration by the museum and consequently its condition deteriorated rapidly. Robert Mikesh of the National Air and Space Museum, who had previously supervised the restoration of Blair's P-51 as part of the museum's collection, became aware of the rapidly declining condition of the VS-44A at Pensacola. He contacted Maureen O'Hara and asked her permission to transfer the aircraft from the U.S. Navy to the New England Air Museum. In 1983, the New England Air Museum received the aircraft from Pensacola on a long-term loan and arranged with the firm of S.C. Loveland to have the aircraft transported from Florida to Bridgeport, Conn.

When Eugene Buckley was appointed President of Sikorsky in 1987 he quickly provided the needed company support to begin the actual restoration. Harry Hleva, a retired product support manager who had worked, as a young man, on the VS-44A project was enlisted to organize a volunteer effort to accomplish the restoration effort. A $150,000 temporary Rubb hangar was erected on Sikorsky property at the Sikorsky Memorial Airport. A dedication ceremony was conducted to kick off the effort and many luminaries from the aircraft's history attended. Mrs. Charles Blair (Maureen O'Hara) and NEAM President Robert Stepanek spoke in support of the restoration project. Dick and Nancy Probert were among the dignitaries in attendance.

For the next ten years, the Rubb hangar, dubbed the "Restoration Hangar" (and referred to, jokingly, as the rubber hangar by Sikorsky workers) was home for the VS-44. During that period of time, a team of approximately 120 volunteers expended close to 200,000 hours accomplishing what has been described as a complete remanufacture of the aircraft. Thirty-five percent of the airframe structure and 97 percent of the aluminum skin of the aircraft were ultimately replaced. Every detail of the aircraft, inside and out, was restored to the configuration delivered to American Export Airlines in 1942. At any given time, it was common to find 35–40 volunteers working on the aircraft. This dedicated, highly talented workforce was comprised primarily of Sikorsky Aircraft retirees, but also included Textron-Lycoming retirees, Platt-Sikorsky Technical School staff and students, airline mechanics, and even a former Luftwaffe officer who was a test pilot for the Me-163 *Komet* rocket fighter.

Representatives from the National Air And Space Museum, including Robert Mikesh, provided technical guidance to the restoration team. Hleva assigned tasks to individual volunteers based on their talents and physical abilities. Many volunteers had worked on this same aircraft when it was originally constructed in 1942. Some volunteers had little aircraft manufacturing experience and were content to help out any way they could. Few aircraft hangars have ever been kept as orderly as the "Restoration Hangar."

It would be impossible to list all the individual contributions of the volunteers. Suffice it to say that the task could not have been accomplished without

The *Excambian* as she arrives at the Sikorsky plant in Bridgeport, Conn. The F-4D Phantom which accompanied the VS-44A on the barge ride from Pensacola was delivered to the New England Air Museum. (Pete Montini-Sikorsky Aircraft)

A group of VS-44A volunteers posed for this 1989 photo. (L-R) Harvey Lippincott, Bill Skelton, Dick Sykes, Stan Beras, Fred Ghirardini, Augie Cuomo, Rudy Opitz, Bob Kretvix, Joe Losardo, Owen Warren, Harry Hleva, Bill Starkey, Charlie Sharp, Ray Reynolds, Art Gustafson, Dmitri "Jimmy"Viner. (Erwin Botsford Photo)

The center section is trial-fitted to the hull in the spring of 1997. (Erwin Botsford Photo)

The restored Radio Operator's panel. (Erwin Botsford Photo)

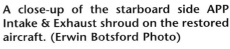

A close-up of the starboard side APP Intake & Exhaust shroud on the restored aircraft. (Erwin Botsford Photo)

A photo taken during the winter of 1996 showing the nearly completed hull next to the completed center section and outer wing panels. (Erwin Botsford Photo)

each and every one of them. It must be noted, however, that a few volunteers, because of their dedication, must be recognized. Stated simply, without Harry Hleva, the job could not have been undertaken or accomplished. Harry's vast knowledge, coupled with absolutely superb leadership qualities, made this restoration possible. Joseph "Joe" Losardo has worked in excess of 18,000 volunteer hours on this project. The late Robert "Bob" Cowell very effectively organized the history of the VS-44s and preserved it by donating it to the Igor I. Sikorsky Historical Archives. Much of the material used in this book was assembled from material accumulated by Bob. Fred

Shubert carries on Bob's tradition and meticulously organizes and catalogues the thousands of pictures, articles, and scraps of paper that chronicle the history of the aircraft and the progress of the restoration team. "Bill" Digney has literally spent years researching and recreating the "original" passenger cabin interior. To say that Bill is a perfectionist is pure understatement. Ken Dineen and John Liddell, both former flight engineers on the VS-44A, provided invaluable consultation, especially relative to restoring the flight deck.

Work progressed steadily and by the spring of 1997 nearly all structural and skin work was complete. The flight deck

had been restored to its original configuration and all control surfaces had been recovered in the proper fabric. On June 18, 1997, the aircraft fuselage was relocated from the VS-44A restoration hangar at the Sikorsky Airport in Stratford, Conn., to the New England Air Museum in Windsor Locks, Conn. Other major components such as the wing center section, wing outer panels, and the engines had been brought to the museum in 1996 and earlier in 1997 as their restorations had been completed. The Sikorsky Company arranged to have the volunteers transported twice weekly to the museum, a distance of 65 miles each way, where, with the help of

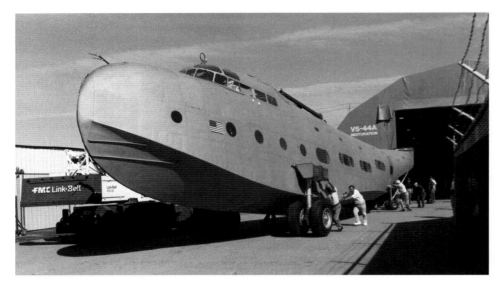

Top: The restored fuselage is moved out of the restoration hangar on June 18, 1997 to begin her trip to the museum. (Erwin Botsford Photo)

Middle: With specially constructed slings and a lot of "know-how," the Meyer riggers turn the hull on its side. Interior bulkheads were reinforced for the trip to the museum. (Erwin Botsford Photo)

Bottom: A good view of the bottom of *Excambian's* hull prior to the journey from Stratford to the museum in Windsor Locks. (Erwin Botsford Photo)

museum staff, they are now completing the assembly of the aircraft. By October 1, 1997, the wing center section had been installed, all four engines and nacelles installed, and the leading edges and control surfaces painted and installed. It is anticipated that the remaining assembly and painting processes will be complete in the summer of 1998. At that time, the aircraft will be moved from the museum restoration hangar to the Harvey Lippincott Civil Aviation Hangar where it will be displayed along with several other significant Sikorsky models.

Another view of the aircraft being nestled onto specially constructed cradles. (Erwin Botsford Photo)

The restored pilots' instrument panel. (Erwin Botsford Photo)

The start of another work day on the VS-44A. Harry Hleva surveys the activity. (Erwin Botsford Photo)

Another photo of the restoration regulars—this one taken in May 1997. (Left to right) First row—Erwin Botsford, Fred Ghirardini, Harry Hleva, John Kopchik. Second row—John Drignet, Vic Politi, Lou Havanich, Joe Losardo, Larry Dzialo, Art Stanko, Frank Dawe, Rudy Opitz. Third row—Bill Smethurst, John Liddell, Joe French, Augie Cuomo, Ken Kraemer, Bob Kretvix, Wally Wanamaker, Pete Peterson, Carl McDonald. (Erwin Botsford Photo)

Left: Four very important figures in the restoration of the *Excambian*. (Left to right) Lloyd Flemming, William B. Meyer, supervisor, John Liddell, VS-44A Flight Engineer, Joe Losardo, who volunteered over 18,000 hours on the restoration effort, and William "Bill" Digney, who supervised the restoration of the passenger cabin and crew quarters.

Former AEA flight engineer Ken Dineen was on the restoration crew.

The End of an Era

The last flying boat designed by the Sikorsky Company was the S-45. The actual design of this aircraft actually started before the design of the S-44 (XPBS-1) was initiated.[19] The S-45 design was iterated many times between 1938 and 1940. A drawing dated March 15, 1938, depicts the S-45 as having a wing span of 236 feet, a one-step hull, a length of 155 feet, six inches, a height of 25 feet, nine inches, and a three-fin tail with a horizontal surface 43 feet wide. The aircraft was designed to carry 100 passengers or a payload of 25,000 lbs., with a range of 5,000 miles and a speed of 200 mph. Perhaps in a "wishful thinking "manner the drawing of the S-45 showed the aircraft with Pan American markings. However, it was not to be. Sikorsky would never again build a flying boat or a fixed wing aircraft. Technological progress during the years of World War II and the construction of hundreds of runways, both in the U.S. and Europe, effectively rendered the large flying boats obsolete. And, of course, Igor Sikorsky was now destined to concentrate on the world's first production helicopter, the R-4.

One question that is frequently asked is "Where are all the flying boats now? While historians and aviation buffs maintain an optimism that there are still some amphibians and flying boats hidden in some long forgotten lagoon or hangar, it is just not likely. Of all the famous Sikorsky amphibians, only two or three non-flyable single engine S-39s, one flyable S-43 in Texas, two partially complete S-43s (one at the Alaska Heritage Museum, one at the Pima Air Museum in Arizona) and Excambian still exist. Of all the S-38s produced no complete examples are known to exist. However, work is nearly complete on an S-38 being manufactured at Born Again Restorations in Minnesota. Buzz Kaplan, owner of BAR, is actually fabricating two S-38s, the first example for Sam Johnson, of S.C. Johnson Co. and the second example to add to his impressive collection of vintage aircraft. BAR is using original S-38 parts wherever possible. Sam Johnson's father was one of Sikorsky's earliest commercial customers and had utilized an S-38 in the late 20s to explore South American rain forests for Carnauba wax trees. Sam plans to retrace the routes of that early voyage in the "new" S-38 named (as was the original)

Carnauba. This magnificent aircraft is expected to fly in mid-1998. No S-40 or S-42 flying boats survive. All of the Boeing 314s are gone as well. The only Martin "boats" left are several "Mars" being utilized as fire-fighters in the Northwest Provinces of Canada. Kermit Weeks has restored and converted a Short Sandringham back into a Short Sunderland. This aircraft, which was previously owned by Charles Blair, is now on public display in Florida. There are numerous "late model" ex-military amphibians, such as the "Albatross" now residing at various air museums, but they just appear too modern to be discussed in the same manner as the big flying boats of the 30s and 40s. Aircraft like the *Excambian* have a romance all of their own which, sadly, will never be recaptured. Thanks to the Proberts, the Blairs, Sikorsky Aircraft, and the work of the restoration volunteers, *Excambian* will live on for many years to come at the New England Air Museum.

Go see her!

Notes

1. The Il'ya Muromets series were the first four-engined bombers ever successfully built and flown.
2. The word "amphibion," derived from "amphibian" was used by Sikorsky to describe his aircraft which were used, not only, in the water and on land like an amphibian, but also in the air—hence "amphibion."
3. The U.S. Navy required a range of 3,450 miles in the patrol configuration and 2,450 miles in the bomber configuration.
4. The XPBS-1 was not the first S-44 designed by Sikorsky. The original design bearing that designation was an eight-passenger amphibian, with a gross weight of 8,900 lbs., designed by Igor Sikorsky and Michael Gluhareff in 1933. That design never entered production and that is the most likely reason for the company re-issuing the S-44 designation.
5. Sikorsky designed a tail gun emplacement for the very successful Russian Il'ya Muromets bomber during World War I. While in use with the Imperial Russian Air Service only one of these bombers was ever shot down.
6. U.S. Navy BuAer press release at the time of issuance of the contract to Sikorsky Aircraft.
7. Lt. Miller went on to accumulate a very impressive war record which included command of the PB4Y-1 Liberator squadron VPB 109. He became one of the highest decorated Navy fliers of WW2.
8. The American Export Vessel *Excalibur* was sunk off the coast of North Africa shortly after the christening of the flying boat named in her honor.
9. VS-44 Mfg. No. 4401 was assigned in 1936 to the XPBS-1.
10. Captain Blair was also employed by Sikorsky as a consulting test pilot during the testing of the VS-44s.
11. *Red Ball In The Sky* by Charles Blair.
12. During war-time service the VS-44As carried anywhere from 16 to 32 passengers, depending on the destination and the nature of the passenger list.
13. In 1942 the Vought-Sikorsky division was separated back into the original divisions of Chance Vought and Sikorsky Aircraft. This was done in recognition of the fact that the helicopter product line now warranted its own dedicated organization.
14. It is interesting to note that Captain Blair, upon arriving in New York, boarded a American Overseas Airline DC-4 and became the first pilot to fly a land-plane non-stop from New York into Shannon, Ireland.
15. Record set by N41881 *Excambian*, captained by Charles Blair.
16. The Antilles Air Boat fleet eventually grew to include 20 Gooses, two Grumman Mallards, four Grumman Albatrosses (which were never used because of FAA certification costs), two PBY Catalinas, and two Short Brothers Sandringham Flying Boats.
17. Shortly after *Excambian* was purchased by Antilles Air Boats, one of the crew lost the bottom half of the forward hatch. Subsequent landings and take-offs allowed untold gallons of corrosive sea-water to enter the hull.
18. Captain Charles Blair died in a crash of a Grumman Goose on September 2, 1978. With ten passengers on board, the left engine exploded and the aircraft impacted the water and sank approximately one mile from the ramp at St. Thomas. Three passengers also perished.
19. As mentioned previously, the S-44 designation was removed from an earlier design and issued to the XPBS-1 somewhat out of sequence.

Ownership Changes of VS-44A *Excambian*

Date	Event
5-May-42	Vought Sikorsky transfers ownership to American Export Airlines
26-Jan-43	American Export Airlines assigns ownership to the U.S. Government
27-Feb-46	War Assets Corporation transfers ownership to Tampico Airlines
3-Apr-47	Tampico transfers ownership to Skyways International Airways
6-Jul-49	Skyways liquidated — Ownership transfered to Seaboard Commercial Finance Corp.
17-Se~50	Ownership transfered from Seaboard to J.L. Boland
6-Dec-50	J.L. Boland transfers ownership to Harry Bomstein
7-Jun-51	Harry Bomstein transfers ownership to Huestis Wells
26-Sep-51	Hugh Wells transfers ownership to Aviation Exchange Corporation
15-Jul-52	Aircraft arrives at Baltimore Harbor Field for refitting
7-Nov-55	Repairs completed at Baltimore Harbor
7-Apr-56	Aircraft being repaired in Lima Peru
14-Jun-57	Aviation Exchange Corp. transfers ownership to Avalon Air Transport
9-Jan-68	Avalon Air Transport transfers ownership to Antilles Air Boats
25-Sep~82	FAA revokes registration number
1976	Antilles Air Boats donates aircraft to Pensacola Naval Air Museum
1983	Aircraft placed on long term loan to New England Air Museum

Bibliography

1. I.I. Sikorsky; *The Story Of The Winged S*, New York, Dodd, Mead & Company, 1967.
2. Archibald Black; *Transport Aviation*, New York, Simmons-Boardman Company, 1926.
3. Charles F. Blair; *Red Ball In The Sky*, New York, Random House, 1969.
4. Frank J. Delear; *Igor Sikorsky, His Three Careers In Aviation*, Dodd, Mead & Co., 1969.
5. Dorothy Cochrane, Von Hardesty, Russell Lee; *The Aviation Careers Of Igor Sikorsky*, Seattle, University Of Washington Press, 1989.
6. *The Beehive*, United Aircraft Corporation, Hartford, Conn.
7. *Air Classics*, Challenge Publications, Inc., Canoga Park, Calif.
8. *Transatlantic Air News*, American Export Airlines, New York, NY.

Comparison of Various Flying Boats

	Sikorsky S-42	Martin M-130	Boeing B-314	Sikorsky VS-44A
Number Built	10	3	12	3
First Flight	5-6-34	9-10-35	1-27-39	1-18-42
Gross Weight, Lbs.	38,000	51,000	82,500	65,000
Empty Weight, Lbs.	19,000	28,000	50,000	32,550
Useful Load, Lbs.	19,000	23,000	32,500	32,450
Load Ratio	100%	82%	65%	99%
Wing Span	118' 2"	130'	152'	124'
Length	69'	90' 10"	106'	76' 3"
Height	21' 9"	24'	27' 7"	27' 7.25"
Top Speed, mph	182	180	192	235
Cruising speed, mph	157	157	183	185
Range, miles	3,000	3,000	3,500	3,500+
Number of Engines	4	4	4	4
Horsepower (each)	700	800	1,200	1,200
Crew	4	5 to7	7 to 10	6 to 12 *
Passengers, Day	32–44	46	74	32
Passengers, Night	—	18–30	36–40	16

Notes:
1. All figures vary with various configurations of each model.
2. * On flights over 8 hours , the crew included:

	2 pilots	2 co-pilots	2 navigators
2 flight engineers	2 radio operators	1 stewardess	1 purser

VS-44A Restoration Volunteers

Sam Agostino
Al Albert
Anvel Agish
Anthony Ambrosino
Charles Ando
Bud Bennett
Gino Berarducci
Stan Beras
John Boran
Erwin Botsford
Jim Cadoret
Charles Canner
Richard Carlson
Ed Carroll
Roy Champagney
Ed Chadbourne
Ted Coley
Bill Conboy
Jeanette Connelly
Bob Cowell
John Cronan
Elmer Crouthers
Augie Cuomo
John Curtiss
Frank Dawe
Pete DiCarlo
Don Dickman
William Digney
Batman & Robin
Bill Dobensky
John Drignet
Larry Dzialo
Charles Farrell
John Fekete
Frank Ferraiuolo
Bob Franklin
Joe French

Greg Fuimara
John Gabor
Freds Ghirardini
John Gill
Robert Goodfellow
Frank Gritsko
Ed Groves
Vince Hardy
Bill Harrington
Phillip Haskell
Lou Havanich
Hary Hleva
Ray Holland
Ken Horsey
Bob Howard
Bob Howey
Rich Jersey
Stu Kerr
Lew King
John Kochiss
John Kopchik
Joseph Kowhacki
Ken Kraemer
Bob Kretvik
Alice Kulhawik
Dan Laezzo
Sigurd Lee
John Liddell
Harvey Lippincott
Bob Lorimer
Joe Losardo
John Marganski
Dennis McCloskey
George McCloskey
Carl McDonald
Peter Mihalko
Leon Montarro
Walter Mottram

Robert Munigle
Ernie Nagel
Rudy Opitz
Bill Pack
Pete Pagliaro
Dom Palumbo
George Paulis
Pete Peterson
Vic Politi
Bob Porter
Jim Rankin
Pete Renson
Ray Reynolds
Melvin Rice
Louis Russo
Thomas Sarkes
Stan Saroka
Ed Sattelberger
Mike Saunders
Robert Shaeffer
Fred Schubert
Charlie Sharp
Bill Smethurst
John Smitonick
Leo Stanejko
Art Stanko
Bill Starkey
Leo Steinerts
Alan Sterling
George Studwell
John Sulinsky
Richard Sykes
Cesare Tomei
John Tullock
Bruce Van Der Mark
John Vavrek
Wally Wanamaker
Henry Wilkinson

Sikorsky VS-44A
Excambian

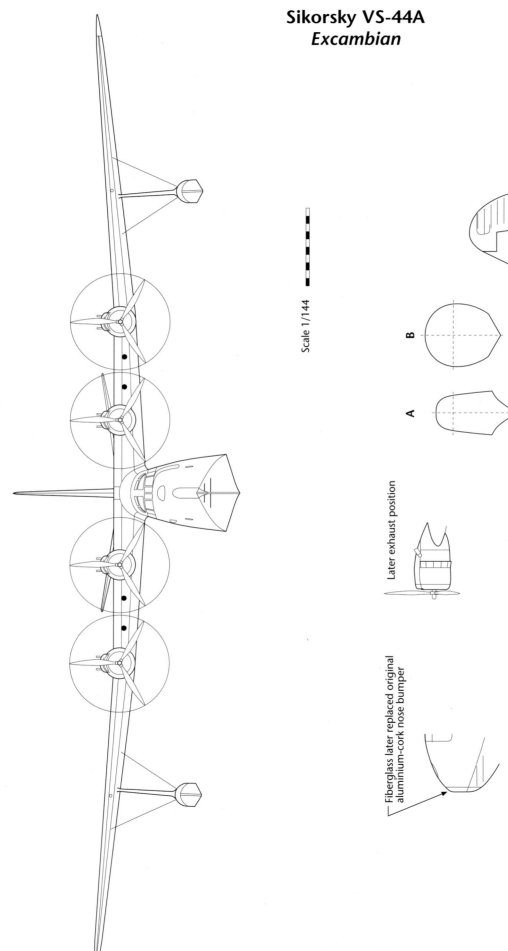

Scale 1/144

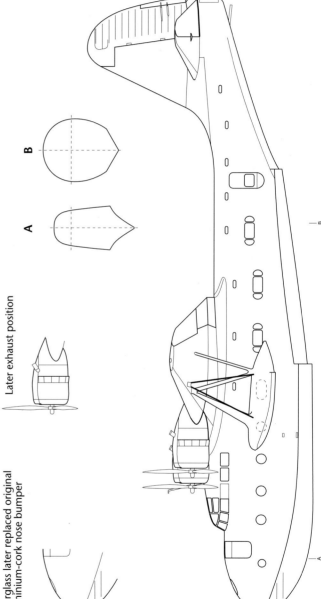

B

A

Later exhaust position

Fiberglass later replaced original
aluminium-cork nose bumper

Sikorsky XPBS-1

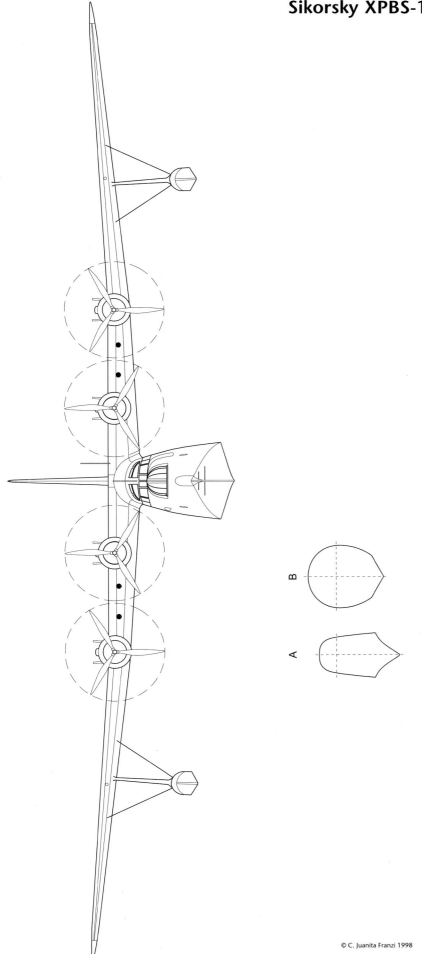

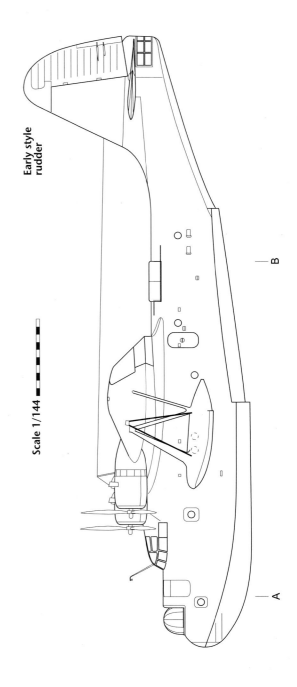

A

B

Early style
rudder

Scale 1/144

Sikorsky VS-44A
Excambian

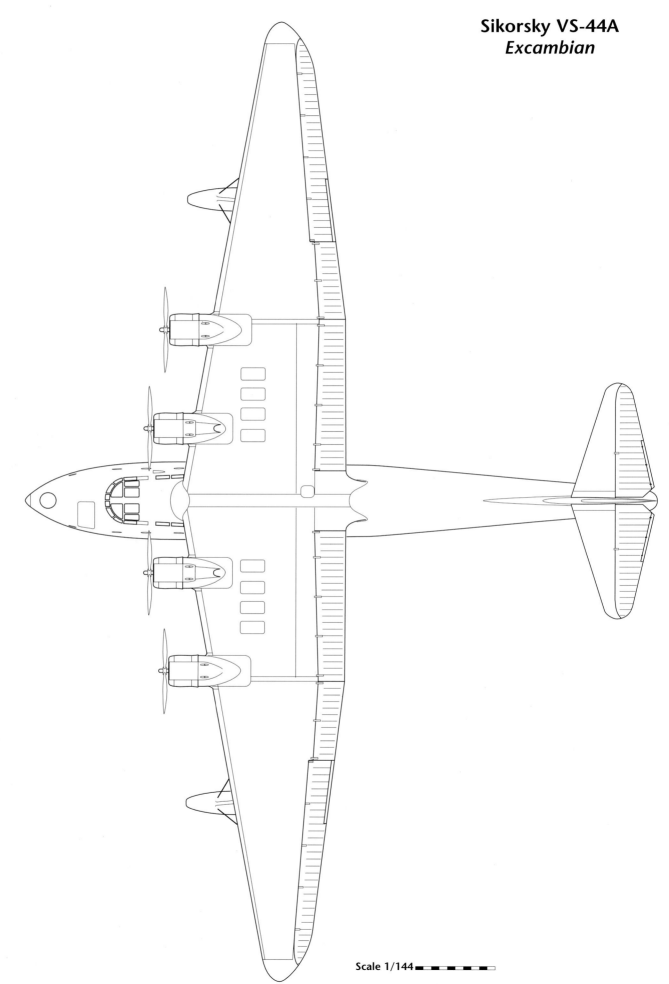

Scale 1/144

Sikorsky XPBS-1

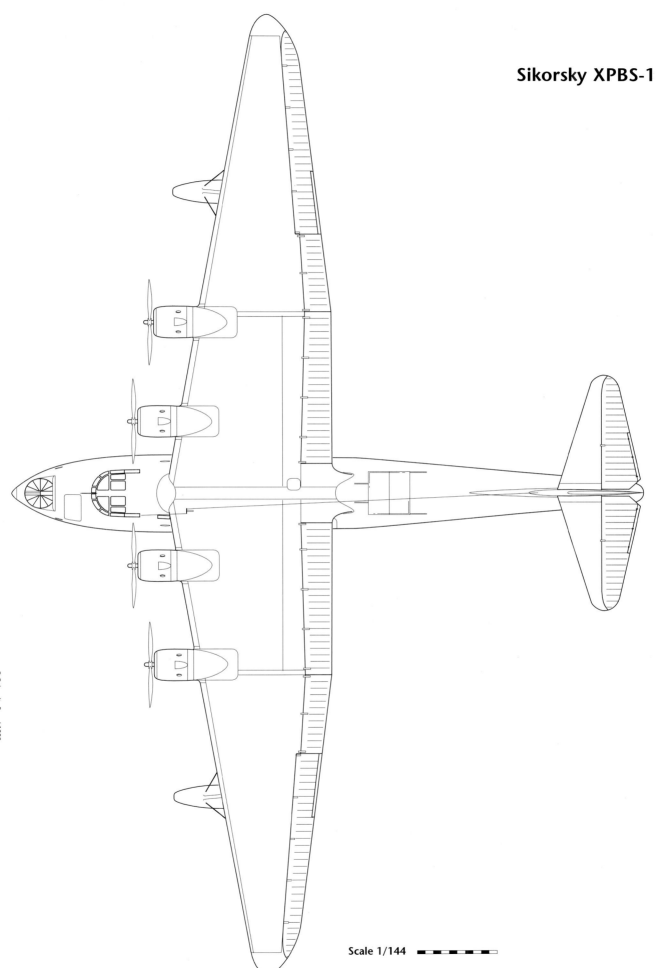

Scale 1/144

Sikorsky VS-44A
Excambian

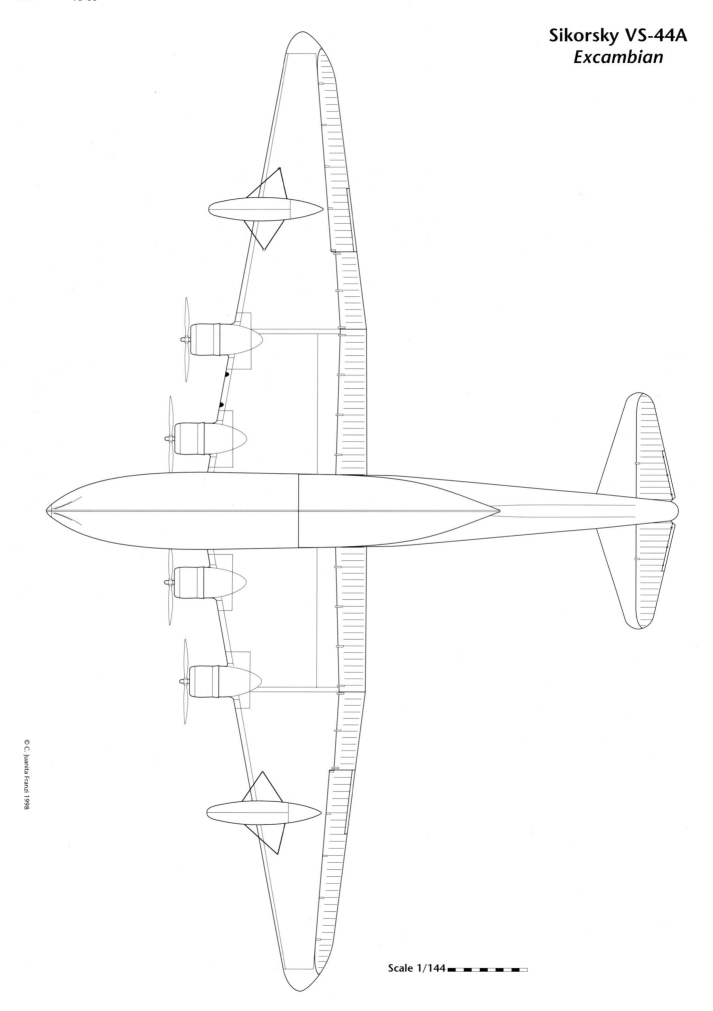

Scale 1/144

Left: An original first day cover issued January 18, 1942, commemorating the first flight of *Excalibur*, the first VS-44A. American Export Airlines became American Overseas Airlines shortly after the end of WWII. AOA was later purchased by Pan American Airlines. Right: Another first day cover, this one commemorating the 50th anniversary of the first flight of *Excambian*, May 11, 1942. Center: A postcard of *Excalibur*. (IISHA)

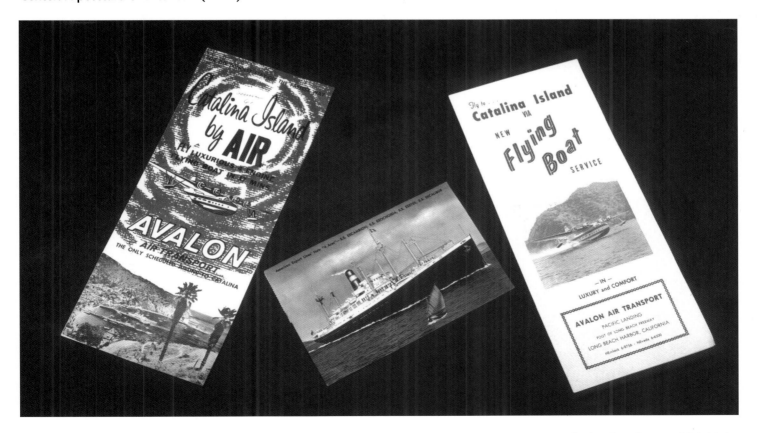

A postcard of the original American Export Lines *Excalibur* framed by two Avalon Air Transport marketing brochures which highlighted the VS-44A. (IISHA)

A REFRIGERATOR AND AN AUTOMOBILE
GO TO WAR !

IT hurtles across the Atlantic between dawn and dinner-time. . .

A huge new cargo-carrying flying boat . . . with squads of soldiers, guns and ammunition . . . a freight-car load of what it takes to smash an enemy.

This, Mr. Hitler — is a picture of a refrigerator and an automobile *going to war*.

Not by ones or twos—but *in fleets*—these ocean-jumping Vought-Sikorskys will be sailing from Nash-Kelvinator assembly lines—ready to fly the fight and might of the U. S. Navy to war.

And when they stretch their wings around this world, there will be proud new Navy *Corsairs* to protect them—new fighting ships that can fly the wings off any Axis 'plane now known!

The *Corsair*, too, carries the colors of Nash-Kelvinator. Its powerful 2,000

h.p. super-charged high-altitude engine is a *quantity* assignment for the men who made last year's automobiles and refrigerators . . . men who have already built thousands of propellers for the Jap-blasting and U-boat-hunting fliers of the United Nations.

This is just a sample, Mr. Hitler, of our 1943 models. Just a picture of what one company is doing—in meeting and

beating a production schedule four times greater than our best peacetime year. And all America's in the fight—buying War Bonds, getting in the scrap metal —in this war *to win*!

So your happy dreams are about over, Adolf — a Nazi nightmare is turning true. The sky is getting darker in the west—the wings of vengeance are coming!

NASH-KELVINATOR CORPORATION

A 1943 Nash-Kelvinator advertisement which included the JR2S-1 they were slated to build. The contract for approximately 200 aircraft was cancelled despite the fact that Nash had begun costruction on a new factory dedicated to building this aircraft.

During the war years when the VS-44As were being constructed, the employment at the Vought-Sikorsky factory increased from 5,000 to 13,000 workers. Housing in the area could not accommodate the sudden, dramatic increase and workers were forced to "hot bunk." The company produced this poster which was hung on the door or in a window to warn delivery men or postmen not to wake the shift workers.

Excambian while serving with Avalon Air Transport. (Dick and Nancy Probert)

Excambian taxiing up to the ramp at Long Beach in her later Avalon Air Transport colors. (Dick and Nancy Probert)

Excambian or "Mother Goose" water taxiing near Avalon. (Dick and Nancy Probert)

Excambian waiting to be moved to an aviation museum. (Charles Blair)

Excambian in service with Antilles Air Boats. (Charles Blair)

Excambian (really *Excalibur*) at Ancon, Peru, showing the floating work platform constructed by Dick Probert to work on the troublesome #2 and #4 engines. (Dick and Nancy Probert)

Excambian in service with Antilles Air Boats. (Charles Blair)

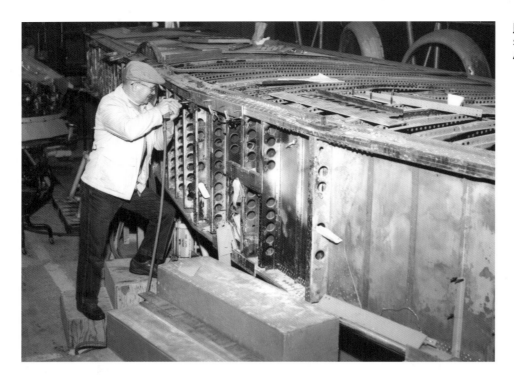

John Kopchik working on the center section. (Pete Montini courtesy Sikorsky Aircraft)

Right: Restoration volunteer Bud Bennett working on the interior of *Excambian*. (Pete Montini courtesy Sikorsky Aircraft)

Left: *Excambian* at New England Air Museum in the fall of 1997 shortly after the four engines were installed. The three original engines that were on the aircraft when it arrived in Connecticut were beyond repair and were replaced with freshly overhauled engines of the same type. The cement pad in front of the NEAM restoration hangar was installed to facilitate moving the big aircraft in and out of the building. (Pete Montini courtesy Sikorsky Aircraft)

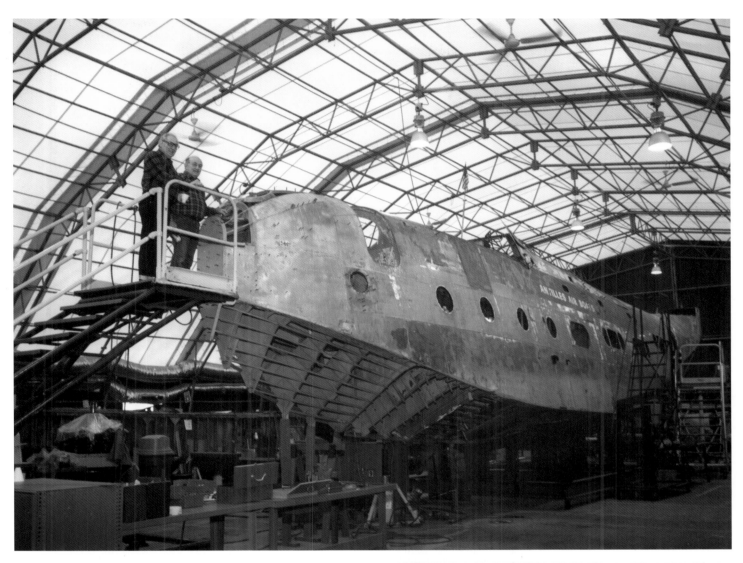

Above: Excambian's hull early in 1988. Tony Ambrosino and Dom Palumbo survey the work ahead. (Pete Montini courtesy Sikorsky Aircraft)

Right: Danny Laezzo, Carl McDonald, and Fred Ghirardini work on the leading edge of an outer wing panel.

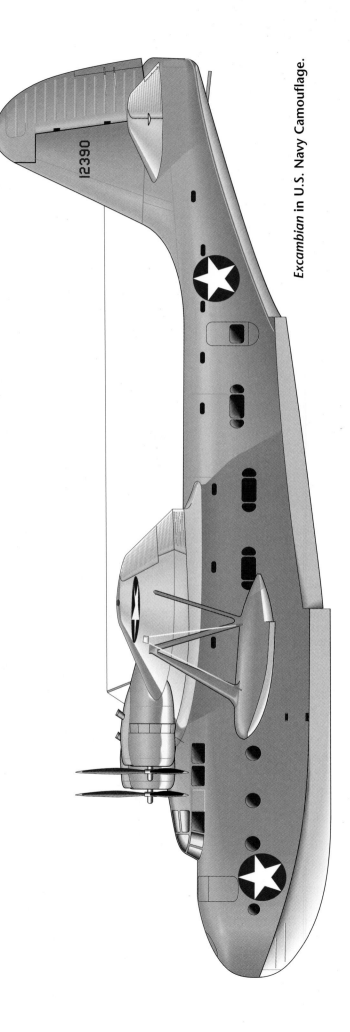

Sikorsky XPBS-1 Patrol Plane.

Excambian in U.S. Navy Camouflage.